Artificial Intelligence and Simulation for Smart Desalination

nova
science publishers
www.novapublishers.com

Water Resource Planning, Development and Management

More information about this series can be found at
https://novapublishers.com/product-category/series/water-resource-planning-development-and-management/

Engineering and Sustainable Development

More information about this series can be found at
https://novapublishers.com/product-category/series/engineering-and-sustainable-development/

Amr Seoudy
Hatem Seoudy

Artificial Intelligence and Simulation for Smart Desalination

nova
science publishers
www.novapublishers.com

DOI: https://doi.org/10.52305/HUXM3650

Library of Congress Cataloging-in-Publication Data

ISBN: 979-8-89530-830-1 (Hardcover)
ISBN: 979-8-89530-944-5 (eBook)

Published by Nova Science Publishers, Inc. † New York

Contents

Chapter 1

Desalination Overview

Abstract

The increasing need for freshwater because of population growth, industrialization, urbanization, and climate change is a major risk to sustainable development globally. Depletion of water resources is exacerbated by pollution, requiring the application of efficient treatment technologies. Seawater desalination and brackish water have become critical water resource solutions, and desalination capacity has been expanding globally. This chapter tells the story of how desalination started and explains the basic steps of the desalination process, which include taking in water, preparing it, and treating it, all of which are important for keeping the water quality good and the system working well. Spotting on various intake systems, as well as pre-treatment methods such as chemical and membrane-based technologies, is critical in optimizing desalination endeavors. Post-treatment processes, such as pH adjustment and remineralization, are discussed to ensure safety as well as stability of the treated water. The paramount importance of improving the outfall system is also discussed to guarantee a stable and eco-friendly system. Moreover, the chapter discusses the economics of desalination related to facilities, such as costs (capital, operating, and energy) and economic concerns like technological innovation and water prices.

Keywords: desalination history, desalination process, seawater intake, pretreatment, post-treatment, effluent outfall, desalination economy

1. Introduction

Recently, it has been increasingly realized that the growing demand for freshwater poses a severe threat to the sustainable development of human societies [1]. There is a plethora of factors, including growing population, expansion in industrialization, urbanization, and climatic change, that have worsened the situation of freshwater resources around the world. In addition, discharges from industrial operations are contaminating natural water supplies, which in turn is causing severe environmental problems. Thus, providing clean water is paramount to satisfying this precious resource's growing demands. With the depletion of freshwater resources, the need for water treatment technologies has risen proportionately. Seawater desalination and wastewater reclamation are widely accepted as solutions to enhance water supplies and counter the problems related to the scarcity of water.

Notably, because nearly 97% of the water resources in the earth are saline, seawater desalination has been considered as a crucial and promising solution for water purification. Recent strides in economically viable seawater desalination methods have significantly improved the desalination process's feasibility as a potential contributor to municipal water supply.

Desalination is seawater, brackish water, tap water, and wastewater effluent treatment process that involves the removal of salt, minerals, and impurities. Thereby yielding clean water suitable for domestic, agricultural, and industrial uses [2].

Accordingly, the global growth in the number of desalination plants has been phenomenal, recording an average annual growth rate of about 6.8% since 2010, which translates into an annual increase of approximately 4.6 million m^3/day in production capacity. In fact, between January 2019 and February 2020, contracts were awarded, and installations of 155 new desalination facilities were installed worldwide, adding another 5.2 million cubic meters a day to the cumulative capacity [3, 4].

The worldwide cumulative capacity for producing freshwater by desalination as of July 2016 was 95.6 million m^3/day, facilitated by 18,983 plants and projects. This cumulative capacity is derived from both existing installation capacity and contracted projects.

As of the end of 2017, considering facilities that have been operational since 1965, the global cumulative capacity for desalination had attained 99.8 million m^3/day, with the effective capacity of the functioning plants projected to represent around 93% of the overall installed capacity.

According to IDRA (International Desalination and Reuse Association), the global cumulative contracted capacity by the year 2024 is 115.01 million m^3/d IDRA+ [5].

2. Desalination History Overview

Desalination has a long and fascinating history that spans thousands of years, with various methods and technologies being developed over time.

Desalination has come a long way, from ancient methods to modern, large-scale plants. As the world continues to face water scarcity challenges, desalination will likely play an increasingly important role in providing freshwater for various uses.

It is crucial to have an overview of the water desalination history to refine our knowledge of it and illustrate a complete vision including causative factors, decisive events, and their effects in the aftermath. The analysis gives advances about what future trends the desalination technologies will be following. All this is fundamental to pre-empting probable obstacles arising from desalination through the informed development of sustainable water governance regimes.

Desalination has been taking place naturally on Earth in the form of seawater evaporation and condensing in the air to form rain. Freezing seawater in Polar and Arctic Regions also creates a vast amount of ice crystals that are made up mostly of pure water, as the salt is effectively barred from the crystallization process [6].

Ancient Civilizations (3000 BCE - 500 CE), covers early methods, Aristotle's contributions, solar evaporation, Egyptian methods, and Roman innovations. Early methods: ancient civilizations, such as the Greeks, Romans, and Phoenicians, used simple methods like boiling, condensation, and solar evaporation to produce freshwater from seawater. Aristotle's contributions: Aristotle (384-322 BCE) wrote about desalination methods, including boiling and filtration. Solar evaporation: people in ancient Middle Eastern and Mediterranean regions used solar evaporation to desalinate water. Egyptian methods: ancient Egyptians used earthenware jars to desalinate water through evaporation and condensation. Roman innovations: Romans developed more advanced desalination techniques, including the use of terra cotta pipes.

Since the Megalithic era, circa 6000 BC, consecutive warm climates of the Mediterranean basin have shaped water resource management and practice developments significantly. One of the exceptional portrayals of the period is the Alcalar funerary monument, Portimão, Portugal. In several phases of Prehistoric periods, from circa 3200 to 1100 BC, there were remarkable developments in hydro-technologies, navigation, and commerce. In particular, the Atlantic Bronze Age that began circa 1300 BC saw a vigorous exchange of copper, tin, and gold between the British Isles, Brittany, and the Iberian Peninsula's western coast. Moreover, Minoan sailors gained supremacy in the Mediterranean Sea and would have used desalination methods by exploiting evaporation through boiling and separating freshwater from the salt as a means to desalinate seawater for drinking [7, 8].

Minoan technologies, including hydraulics and navigation, were exported to the mainland in later periods of the Greek civilizations in the Mycenaean, Archaic, Classical, Hellenistic, and Roman periods [9].

In Roman and Hellenistic times, there were great strides in the fields of hydraulics and navigation, with the construction and utilization of big seafaring ships. The ancient Greeks are noted by history to be the first to formulate philosophical concepts of the inherent nature of water. Thales of Miletus (c. 624–548 BC), a great exponent of the School of Ionia and one of the seven sacred sages of ancient Greece, postulated that water is an elemental substance, fertile as well as versatile. Furthermore, Thales continued to expound the theory that seawater encompasses the great ocean that encircles the Earth, which was considered by him to be the primal material foundation of all forms of life on Earth [10].

In the early period of philosophical reflection, Empedocles (c. 490–430 BC), a philosopher from Sicily, had proposed a theoretic system in which he postulated the concept that the world consists of four indispensable elements, namely earth, water, air, and fire. Contemporary treatment of these groups of elements could easily be reformulated as the fundamental factors to human health, i.e., water, land, energy, and atmosphere.

Historical accounts show that in about 750 BC, the Phoenicians, followed by the Carthaginians—sea-going merchants of modern-day Lebanon and Tunisia—began to be noticed on Portuguese shores. The era likely signaled the start of irrigation methods, in addition to the invention of the waterwheel Nora and the Shaduf, an early hand-powered machine to lift water. Furthermore, there are also documented interactions with Greek sailors along the Iberian coastline from around 630 to 535 BC.

The idea and desalination technology has gained recognition in various phases of history. Particularly during late Archaic Greece, Ionian thinkers based in Miletus understood the possibility of reuse of freshwater. Similarly, during the Warring States period (around 475–221 BC) in China, the Classic of Mountains and Seas documented the application of bamboo mats during rice steaming that were found to yield a thin outer layer efficient in salt adsorption and ion exchange.

Moreover, the ancient Greek philosopher Aristotle (384–322 BC) refined the concept of the hydrological cycle and energy transformation in his works. He believed that though saltwater can be converted to freshwater by vaporization, the water produced will not revert to its salty state when the condensation process takes place. He also observed that a container of wax, when immersed in seawater over a long time, actually strained off salt, thereby producing drinking water. For example, in the time of approximately 200 AD, Alexander of Aphrodisias quoted the desalination method employed by seamen, who boiled seawater to generate steam, which was absorbed by sponges and hence yielded potable water [7, 8, 11-13].

Middle Ages (500-1500 CE), covers limited advancements, and Arabian innovations. Limited advancements: during the Middle Ages, desalination techniques saw limited advancements, with most methods still relying on simple boiling and condensation. Arabian innovations: Arabian alchemists developed new methods, including the use of alembics for distillation.

Before the Industrial Revolution, desalination was mostly used in ships that needed freshwater. Thomas Jefferson developed another way of desalinating seawater using heat, which was later implemented in the United States. Desalination methods have been advancing since the 1500s. Educational stickers that outlined Jefferson's process were made, distributed, and placed on the back of sailing permits for all ships in the United States. powered desalination. Distillation: The Industrial Revolution brought significant advancements in desalination, with the development of distillation methods. Distillation became a widely used method for desalination. Steam-powered desalination: steam-powered desalination plants emerged, enabling larger-scale production of freshwater [13].

Industrial Revolution (1500-1800 CE), covers distillation, and steam-powered desalination. Distillation: The Industrial Revolution brought significant advancements in desalination, with the development of distillation methods. Distillation became a widely used method for desalination. Steam-powered desalination: steam-powered desalination plants emerged, enabling larger-scale production of freshwater.

Modern Era (1800-1950 CE), covers distillation, multi-stage flash distillation, and reverse osmosis. Distillation: in the 19th century, distillation became a common method for desalination, particularly for ships at sea. Multi-stage flash distillation: the multi-stage flash distillation (MSF) process was developed, allowing for more efficient and larger-scale desalination. Reverse Osmosis: the concept of reverse osmosis (RO) was first observed by Jean-Antoine Nollet in 1748, but it wasn't until the mid-20th century that RO membranes were developed, revolutionizing desalination.

Contemporary Era (1950 CE – 2020 CE), covers large-scale desalination, technological advancements, and renewable energy integration. Large-Scale Desalination: Large-scale desalination plants began to emerge in the Middle East and other water-scarce regions, after World War II. Technological Advancements: Improvements in materials, design, and energy recovery have increased efficiency and reduced the costs of desalination. Renewable Energy Integration: There's a growing focus on integrating renewable energy sources, such as solar and wind power, into desalination plants.

Large-scale desalination plants only became technologically viable after World War II [14, 15]. Nevertheless, there are various historical records that demonstrate the experimentation and research of desalination technologies by ancient civilizations [6, 10]. After World War II, the United States conducted extensive research to establish desalination technologies. In this regard, the U.S. Department of the Interior created the Office of Saline Water as a part of its operations in saline water conversion. In 1974, it was merged into the Office of Research of Water Resources. The investigations also took place at several universities, followed by development at different chemical companies [16]. The development of potable water production has come a long way, whereby solutions with broad applicability have been realized [17]. With the growing world population and the subsequent escalation of freshwater demand, interest has been renewed among investors for the realization of technologies for potable water production in remote areas, especially sea vessels [18]. As a result, various technologies for the simplification of desalination processes have been reported in recent literature [19, 20].

Contemporary Developments (2020 CE – Future), covers renewable energy integration, innovative technologies, artificial inelegant, sustainability, and global cooperation. Renewable Energy Integration: There's a growing focus on integrating renewable energy sources, such as solar and wind power, into desalination plants. Innovative Technologies: Researchers are exploring new technologies, like graphene-based membranes and advanced materials, to further improve desalination efficiency and sustainability. Artificial Inelegant: For economic and smart desalination. Sustainability: Focus on reducing environmental impacts and increasing sustainability. Global Cooperation: International cooperation to address water scarcity challenges.

3. Desalination Process Overview

As we mentioned previously, the Desalination Process is complex and involves purifying seawater or brackish water by removing salt, minerals, and other pollutants to produce freshwater. Desalination systems use different technologies to produce freshwater of varying quality from different sources of raw water. Regardless of the technology used, this process typically involves several stages, including the intake stage, pretreatment stage, desalination stage, post-treatment stage, and outfall stage, as shown in Figure 1.

3.1. Intake Stage

Desalination facilities require an intake system that is reliable and consistent, one that can supply an adequate quantity of seawater necessary to achieve production objectives. While this basic need appears simple, it is complicated by the variable nature of the water source, one that undergoes continuous changes. Parameters such as heavy waves and shifting currents can negatively impact structural stability and water level and dramatically alter both the quality and temperature of seawater.

Notably, these oscillations often grow more intense as you approach the coastline, occurring with increased frequency and amplitude. The operational issues are added to by the corrosive nature of seawater and the potential for marine life to foul equipment and systems. To satisfy the design criteria as stipulated, there is a need to conduct a thorough evaluation of the conditions at the intake point. Along with physical features, meteorological and hydrographic data, marine biological factors, and the possible effects of fouling, pollution, and navigation traffic, this analysis needs to look at everything. It is only through this careful analysis that a suitable intake design can be established.

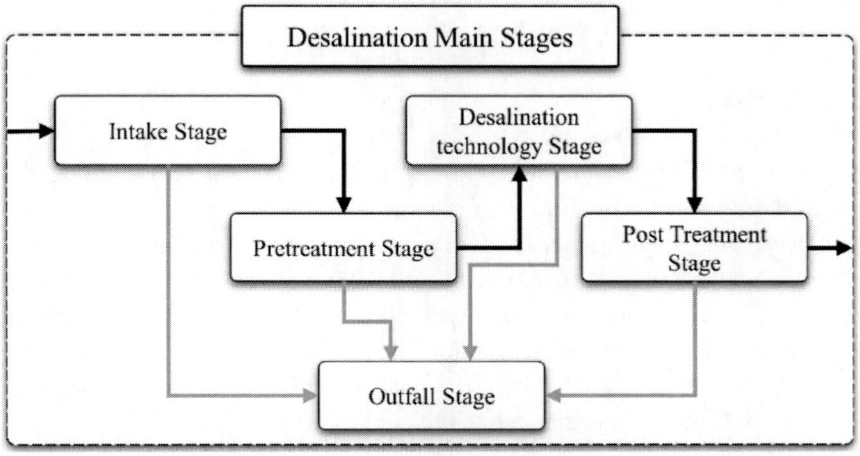

Figure 1. Desalination System Main Unites.

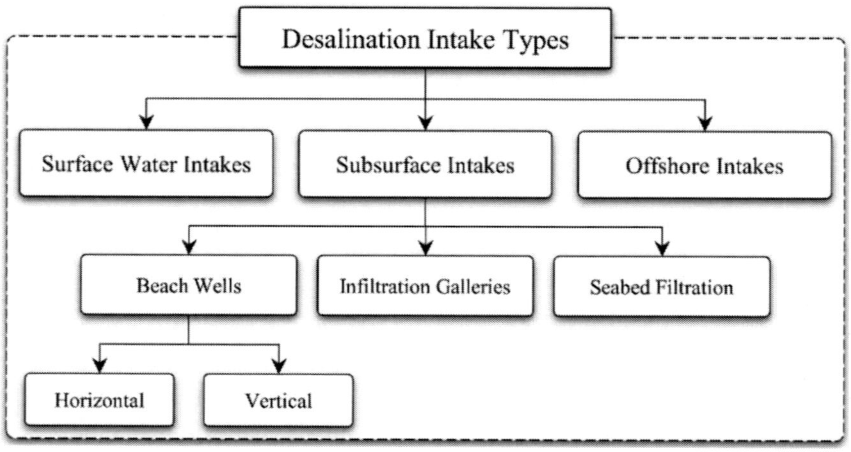

Figure 2. Desalination System Intake Types.

Historically, seawater intake systems represented some 4÷12% of the overall capital cost that is specific to desalination plants. However, according to current design trends, the intake systems make up at least 35% of the total project capital costs. This means that problems with the intake mechanisms can have a big effect on how well the desalination plant works as a whole. Desalination system intake types can be classified into three main types, surface water intakes, offshore intakes, and subsurface intakes, as shown in Figure 2.

Surface Water Intakes: Seawater desalination plants have typically employed open-sea surface water intakes with mechanically cleaned screens. These plants operate in a similar way to electric power plants, which use the same type of intake system to take in condenser cooling water. The infrastructure typically consists of a pump station and screening chamber on land, with direct access to the open sea through a concrete channel, jetty, or intake pipe extending several hundred meters into the marine environment.

Offshore Intakes: Offshore submerged intakes are the most popular design for seawater intake, especially in shallow coastal waters. The intakes are used by power plants and desalination plants to take in a greater and more reliable quality of water, which is less likely to cause plant shutdowns due to storm activity, algal blooms, and jellyfish infestations. Careful placement and design of an offshore intake can also help protect the environment by reducing the number of marine animals that come into contact with and are sucked into the water [21].

Subsurface Intakes: These are the different types of subsurface intakes that are used to get water from below the seabed. They can be horizontal or vertical beach wells, infiltration galleries, or seabed filtration schemes. These architectural designs all function through the separation of open seawater from the intake location via the intervention of a geological structure [22, 23]. The use of subsurface intakes is feasible where the geology beneath the surface is suitable for extracting water while, at the same time, providing some natural filtration [24].

3.2. Pre-Treatment Stage

Pre-treatment of feed water is necessary for removing turbidity and minimizing the levels of microorganisms, colloids, dissolved organic and inorganic substances, total dissolved solids (TDS), silt density index (SDI), and a multitude of pollutants to levels acceptable for downstream treatment processes. Pre-treatment systems currently available in the market can be generally categorized as conventional and membrane-based systems, as described in Figure 3.

Conventional pretreatment techniques, used in seawater desalination are widely categorized into two groups: chemical and physical pretreatment. Chemical pretreatment, on the other hand, involves a chemical addition of scale inhibitors, coagulants, and disinfectants. Coagulation and Flocculation: Chemicals are added to remove dirt and other suspended particles from the water and to achieve the elimination of turbidity. Sedimentation: The water is left to sit, allowing the particles to settle to the bottom.

Flotation: is a critical process for the elimination of particulate organic impurities such as algal cells, grease, and oil that cannot be successfully eliminated by sedimentation and can cause serious problems to downstream operations. Disinfection: disinfection purpose is to prevent any biological growth responsible for causing clogging of filters and membranes.

De-chlorination: Removal of residual chlorine in feed water especially in reverse osmosis process to avoid membrane damage through oxidation Scaling Prevention: Chemicals are added to prevent scaling on equipment. Physical pretreatment is a mechanical screening operation that attempts to eliminate particulate matter through the use of screens and filters. Screening: consists of a mesh kept at an inclination to the stream of raw water to prevent the system and debris in the raw water are stopped at the grids which are removed by a mechanical rake attached to the screens. Filtration: is a process that assists in the elimination of suspended solids from water as it flows through filter medium.

Membrane based systems, include microfiltration, ultrafiltration, and nanofiltration. Microfiltration (MF): MF is conducted using membranes with pore sizes between 10 µm and 0.1 µm, which enables effective removal of colloidal substances, suspended particulate matter, and, to a lesser degree.

Bacterial contaminants. Ultrafiltration (UF): UF membranes have pore sizes ranging from 0.01 to 0.1 µm and thus allow efficient removal of suspended solids, colloidal material, silt, bacteria, and viruses from the source water. Nanofiltration (NF): NF is a relative membrane filtration process used most commonly for the treatment of waters with relatively low total concentration of dissolved solids. NF membrane pore diameters range from 1 to 10 nanometers and thus fall between ultrafiltration and RO membranes in rejection capacity.

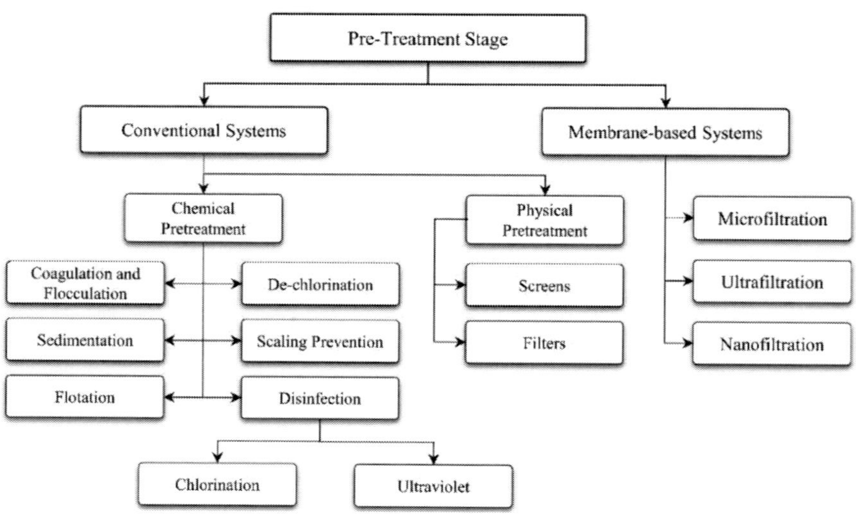

Figure 3. Pre-treatment Stage Main Unites.

3.3. Desalination Stage

Desalination technologies for water are generally divided into two main groups: Phase change or thermal processes and non-phase change processes through the utilization of membranes [25-27].

Phase change, or the thermal processes, functions based on evaporating seawater by contact with a heat interface, the evaporation surface. The dissolved salts are left behind during this process. The vaporized freshwater is then condensed in pipes that cool it down, delivering high-pressure, high-quality freshwater that is free of salt [28]. The four main types of thermal desalination used in business are multi-effect distillation (MED), multi-stage flash distillation (MSF), Mechanical Vapor Compression (MVC), and Thermal Vapor Compression (TVC) [25, 29, 30]. Besides, solar energy is utilized by several technologies as the power source, e.g., Freeze Distillation (FD), Gas Hydrates (GH), Solar Stills (SS), Solar Chimney (SC), Adsorption Desalination (AD), and Humidification & Dehumidification (HDH) systems; however, these technologies are not yet widely commercialized.

Non-phase-change desalination technologies, on the other hand, use membranes or materials with similar properties to physically separate dissolved salts from water by pressing them apart directly. This requires a lot of energy or pressure. Membrane processes are Microfiltration (MF), Ultrafiltration (UF), Nanofiltration (NF), Membrane Bioreactors (MB), Membrane Distillation (MD), Electrodialysis (ED), and Reverse Osmosis (RO) [31]. The processes are pressure-driven mechanisms by nature that enable particles, bacteria, and salts to be removed from water by size exclusion via membranes of varying pore sizes [32], Table 1, show desalination process classifications based on phase change including hybrid systems.

3.4. Post-Treatment Stage

Desalination systems production must be treated before storage and transmission to the consumers. The treatment is required for various reasons, including the ability of the product water to lead to excessive corrosion of the pipeline transmission systems. The membrane-produced water usually needs to be adjusted in pH and calcium salts added, i.e., calcium carbonate or sodium carbonate/bicarbonate, to stabilize it. In addition, the product water needs to be disinfected to prevent bacterial growth prior to its introduction into the distribution system. The most common needed treatment processes are remineralization, disinfection, and final pH adjustment.

Remineralization: Minerals are added to the freshwater to improve its taste and quality; the remineralization post-treatment process is targeted at raising the carbonate hardness of desalinated water through the regulation of calcium and alkalinity levels to pre-established target levels. This is done to reduce the corrosiveness of the water, ensure the necessary Sodium Adsorption Ratio (SAR) for agricultural irrigation, and ensure appropriate salinity and hardness according to potable water health safety standards. Disinfection: to remove any bacteria or other microorganisms. It is not adequate to solely depend on the membrane process to be able to guarantee overall disinfection of the product water. In addition, recontamination of the RO product water can occur during the post-treatment stage, mainly as a result of the introduction of microorganisms related to chemical reaction products used in this phase, and also from atmospheric exposure during filling and emptying operations of intermediate and buffer storage before distribution. In order to properly protect the drinking water generated by the RO system, it is necessary to carry out disinfection during the conditioning phase of post-treatment.

Final pH adjustment: pH value is an important parameter for determining the quality of water after treatment processes. The ideal pH value tends to be determined by treatment concerning alkalinity and calcium concentration. This relation arises from the fact that there are two underlying factors: firstly, the ability of water to dissolve or precipitate calcium carbonate is regarded as a key indicator of water stability; secondly, alkalinity, calcium concentration, and pH are the three principal factors known to affect this dissolution or precipitation ability. Storage and distribution: The freshwater is stored and distributed to consumers.

Table 1. Desalination process classifications based on phase change

Desalination Technologies	
Phase Change	
Multi-stage Flash (MSF)	Gas Hydrates (GH)
Multi Effect Distillation (MED)	Solar Stills (SS)
Vapor Compression (VC)	Solar Chimney (SC)
Membrane Distillation (MD)	Adsorption Desalination (AD)
Freeze Distillation (FD)	Humidification & Dehumidification (HDH)
No Phase Change	
Reverse Osmosis (RO)	Forward Osmosis (FO)
Nanofiltration (NF)	Electrodialysis (ED)
Ion Exchange (IE)	Capacitive Deionization (CDI)
Hybrid Systems	
Forward Osmosis / Nanofiltration (FO-NF)	Electrodialysis / Reverse Osmosis (ED-RO)
Reverse Osmosis / Membrane Distillation (RO-MD)	Forward Osmosis / Membrane Distillation (FO-MD)
Forward Osmosis / Electrodialysis (FO-ED)	Freeze Distillation / Membrane Distillation (FD-MD)

3.5. Outfalls Stage

In desalination, the production of high-salinity brine is a concern as far as its discharge into receiving waters with minimal environmental effect is concerned. Brine discharge is currently a main issue in countries employing desalination technologies to generate freshwater. Brine is usually released as a turbulent jet with a much higher initial density compared to the ambient receiving water. To reduce environmental impacts, therefore, quick mixing of the released brine is desirable, with an understanding of the dynamics of jet formation.

Because the jet is denser than the surrounding water, it floats less than it weighs and settles on the bottom of the sea at a certain distance from where it started, depending on its initial speed, buoyancy, discharge angle, and bathymetry. Once it reaches the seabed, the jet is a low mixing rate gravity current, thus emphasizing the need to ensure maximum dilution while the jet proceeds along the water column [33-35]. The expected effects of a desalination plant on the environment depend on the specifics of the design, such as the pretreatment and discharge systems, as well as the location of the facility, such as where the water comes in and goes out. The concentrate might still have corrosion from the desalination unit and chemicals that were used for pretreatment, stopping scale buildup, or cleaning.

It is necessary to choose the intended discharge setup based on a few considerations: To provide for the formation of a stable seabed that will be little influenced by diurnal currents and not likely to be drastically changed under storm conditions. It is important to maintain an appropriate distance from the intake to minimize the risk of concentrate recirculation. The goal is to reduce the effect of wave energy on the structural integrity of the discharge system to acceptable levels. To have sufficient water depth so that the discharge is contained within the water column and will not surface. Also, there needs to be clear language about how much dilution is needed to create a mixing zone that is contained and always has salinity levels that are acceptable and dissolved oxygen levels that are close to background levels. Any chemicals remaining from treatment have to be well dispersed away from the outfall site for environmental protection.

The traditional disposal options such as surface-water discharge, discharge into the sewer system, deep well (subsurface) injection (high-capacity injection wells, shallow injection wells, horizontal well injection), evaporation ponds, and mechanical evaporation [36]. Desalination system outfall types are shown in Figure 4.

Surface-water discharge: is a technique used to dispose of brine through direct injection of brine into water bodies such as seas, rivers, bays, lakes, and other open water bodies. It should be observed that brine release can have adverse impacts on marine ecosystems mainly as a result of high salinity levels as well as the possibility of introducing chemicals that would not naturally occur in the surrounding aquatic environment [37, 38].

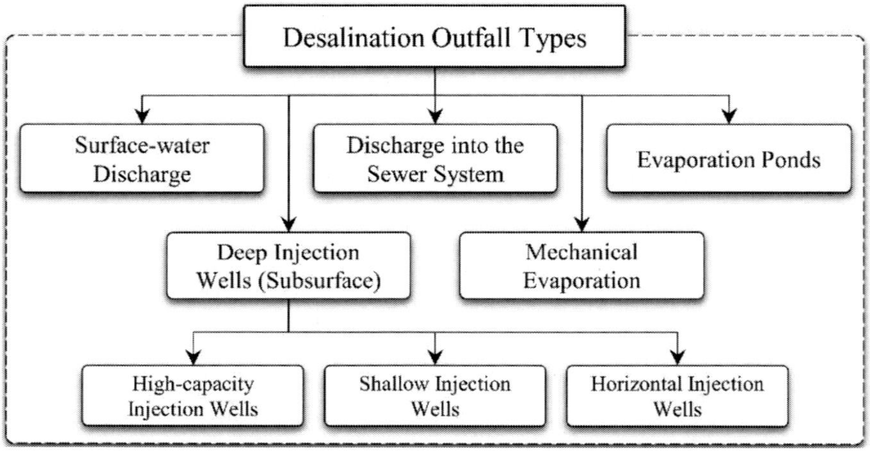

Figure 4. Desalination System Outfall Types.

Discharge into sewer system: The discharge of wastewater is one of the brine disposal methods, where the injection of brine occurs into nearby wastewater collection systems. This practice is largely followed by small Brackish Water (BW) desalination plants because of the high levels of Total Dissolved Solids (TDS) in the brine, which can negatively impact the functionality of the receiving wastewater treatment plant (WWTP). Excessive salinity levels; can inhibit the biologic treatment processes in a WWTP, particularly when the TDS content of the influent is more than 3,000 mg/L. As the TDS content of seawater (SW) brine can exceed 55,000 mg/L, it is required that the WWTP capacity must be at least 20-fold the volume of brine discharge per day to maintain the influent TDS level below the harmful concentration of 3,000 mg/L. Further, high salinity of the final effluent can lead to both environmental and regulatory problems during disposal. Hence, sewer discharge as a waste disposal technique is predominantly coupled to brine water desalination facilities and is scarcely undertaken with regard to seawater desalination [39-42].

Deep well (subsurface) injection: The primary goal of a well-designed discharge system is to minimize the extent of the near field while, concurrently, ensuring that environmental impacts within this area are kept to a minimum. A second goal is to make sure that the discharge is well mixed so that it doesn't have any big effects on faraway areas and doesn't go back to the general outfall area or plant intake system to be used again. There are many types of deep well (subsurface) injections such as; high-capacity injection wells, shallow injection wells, and horizontal well injection.

Evaporation ponds: In areas of hot, dry climates, solar evaporation ponds are used as a disposal for concentrated wastes. It is a process that entails the use of large tracts of land, apart from the capital investment needed in lining the ponds effectively to avert the possibility of brine emission into the neighboring groundwater systems. Aside from that, after evaporation has been completed, the salt sludge generated has to be taken out and hauled to a permitted landfill, thus residual management becomes a key component of the overall cost structure. Research on an experimental scale is also in progress on the disposal of brine or concentrated waters by the use of solar ponds coupled with various evaporator systems. These emerging techniques can be used as effective alternatives for the disposal of concentrated wastes in the future [43].

Recycling into the feedwater stream: Recent breakthroughs in recycling material concentration techniques are now being implemented at specialty processing plants. Specifically, one novel approach includes mixing nanofiltration concentrate with high-salinity aquifer water and then passing this through seawater membranes. The mixing of the concentration with the saline groundwater is expected to improve the recovery efficiencies simultaneously while allowing significant reuse of the concentrate stream. In addition, many instances have been reported where low TDS concentrates are blended into the influence of municipal wastewater treatment plants. Although this is an effective disposal alternative, it generates costs associated with the volume of effluent to be treated. It can also require modifications to the wastewater treatment processes to manage the unique water chemistry of the concentrate [22].

4. Desalination Economics Overview

Desalination economics involves evaluating the costs and benefits of desalination projects. The economics of desalination are influenced by various factors, including: cost components, economic factors, financing options, economic benefits, economic challenges, economic evaluation, and economic future directions and opportunities [44].

Cost components include capital costs, operating costs, and energy costs. Capital costs: the initial investment required to build a desalination plant, including costs for land acquisition, design, construction, and equipment. Operating costs: the ongoing expenses associated with running a desalination plant, including energy, chemicals, maintenance, labor, and other ongoing expenses. Energy costs: the cost of energy required to power the desalination process, which can be a significant portion of operating costs, influenced by energy source and efficiency [45, 46].

Economic factors, include water pricing, economies of scale, and technology. Water pricing: the cost of desalinated water, influenced by production costs, market demand, and regulatory frameworks. Economies of scale: larger plants can benefit from economies of scale, reducing costs per unit of water produced. Technology: advances in desalination technology can improve efficiency, reduce costs, and increase sustainability [46].

Financing options, include public funding, private investment, and grants and subsidies. Public funding: governments and public entities can provide funding for desalination projects. Private investment: private companies can invest in desalination projects, often through public-private partnerships. Grants and subsidies: governments and organizations can provide grants and subsidies to support desalination projects [47].

Economic benefits, include water supply, water security, economic growth, and job creation. Water supply: desalination provides a reliable source of freshwater, which can support economic development, agriculture, and industry. Water security: desalination can enhance water security by diversifying water sources and reducing dependence on traditional sources. Economic growth: desalination can support economic growth by providing a reliable source of water for industries, agriculture, and municipalities. Job creation: desalination projects can create jobs during construction and operation [48].

Economic challenges include high costs, energy consumption, and environmental impacts. High costs: desalination can be expensive, particularly for small-scale plants or plants located in remote areas. Energy consumption: desalination is energy-intensive, which can lead to high operating costs and greenhouse gas emissions. Environmental impacts: desalination can have environmental impacts, such as harm to marine life and increased greenhouse gas emissions [49].

Economic evaluation, includes cost-benefit analysis, economic modeling, and sensitivity analysis. Cost-benefit analysis: evaluating the costs and benefits of desalination projects to determine their economic viability. Economic modeling: using economic models to forecast the economic impacts of desalination projects. Sensitivity analysis: analyzing the sensitivity of economic outcomes to changes in key variables, such as energy costs or water demand [50, 51].

Economic future directions and opportunities, include renewable energy integration, innovative technologies, public-private partnerships, and sustainable financing. Renewable energy integration: integrating renewable energy sources, such as solar or wind power, can reduce energy costs and environmental impacts. Innovative technologies: developing new technologies and materials can improve desalination efficiency, reduce costs, and increase sustainability. Public-private partnerships: collaborations between public and private entities can help finance and deliver desalination projects. Sustainable financing: exploring innovative financing options and public-private partnerships to support desalination projects.

Desalination economics is a complex field that requires careful consideration of various factors, including costs, benefits, and environmental impacts. By understanding the economics of desalination, stakeholders can make informed decisions about desalination projects and work towards more sustainable and cost-effective solutions.

Conclusion

The history of desalination reflects growing human efforts to address water scarcity, which have progressed from the ancient era to advanced technological ones. With extensive water shortages, it is essential to learn from previous innovations to establish sustainable desalination approaches. Desalination as a process consists of major stages: intake, pre-treatment, desalination, post-treatment, and outfall play an integral role in securing clean, safe, and sustainable water production. Various technological approaches, such as thermal technologies and membrane processes, are available to fulfill freshwater demands. Nevertheless, Reverse Osmosis is the most popular choice. It is important to ensure the quality and safety of desalinated water through efficient post-treatment processes to avoid any possible systemic problems. Additionally, the brine discharge problem needs paramount solutions in terms of safeguarding marine ecosystems, thereby rendering ecological considerations a major consideration in the planning. The economics of desalination should be dealt with optimum feasibility. Proper analysis of expenses and financing, especially regarding energy consumption, is crucial to the viability of desalination in the future. The new technologies, especially the incorporation of renewable energies, are a promising solution; we can enhance our water security while enabling a sustainable economy. The way forward lies in a concerted effort that balances technological development, environmental sustainability, and economic viability.

Chapter 2

Desalination Challenges and Sustainability

Abstract

Water desalination is becoming increasingly important in meeting the rising need for freshwater, which is being driven by population expansion and climate change. However, there are numerous challenges facing desalination including financial obstacles linked to huge capital outlays for both plant capital expenditure (CapEx) and operating expense (OpEx), Negative Environmental Impact Challenges that raise questions about its ecological sustainability; technical complexities related to efficiency and optimal operating conditions, governmental regulations; including planning and implementation of desalination, Federal and environmental laws, Taxes and Customs that complicate the execution of projects; in addition to social and psychological challenges. This chapter highlights these challenges and suggests activities and actions towards sustainable and affordable Desalination. These include the financial model, education, training, technical studies, optimization of design, economics, development and research, localization, social and psychological awareness, and public-private partnerships (PPP).

Keywords: desalination challenges, desalination sustainability, public-private partnerships, social challenges, psychological challenges

1. Introduction

Water desalination technologies have emerged as a critical area of research in response to the increasing global demand for freshwater, driven by population growth and climate change. Water security is one of the greatest challenges for development plans of our time. Dramatically, population growth and climate change are putting increasing stress on finite water resources in some areas, especially in lower-middle-income countries. Relying on various and integrated water sources such as; freshwater rivers and lakes, groundwater, rainwater, reuse of agricultural water, reuse of treated wastewater, and desalination, has become essential to overcome water scarcity. Water scarcity, caused by non-integrated water resource management, pollution, climate change, and population growth, is the most urgent problem. It has severely affected biological diversity, threatening more than two-thirds of the natural habitat[52]. Thus, in this chapter, a brief discussion on the challenges facing desalination to be a sustainable source for supplying freshwater from different prospective is introduced with proposed approaches to be considered to overcome these challenges and achieve sustainable desalination.

2. Desalination Challenges

Desalination facilities have several obstacles that limit operational efficiency and cost-effectiveness, thereby jeopardizing their longevity and sustainability. Most recent research and studies have shown that there are four main challenges facing the water desalination industry: financial challenges, negative environmental impact challenges, technical complexity challenges, and government regulations and permissions. Countries that depend on natural water sources such as freshwater rivers, lakes, and rains to secure freshwater, meanwhile having lower-middle-income economies, face different challenges using desalination as a new source to secure freshwater, in addition to the challenges countries mainly depend on desalination as a main secure to secure freshwater faces. In this chapter, these challenges will be addressed, considering some sub-points that may not been discussed previously, including exposure to some of the basic challenges of water desalination plants, which have a direct impact on this industry, especially from a social and economic point of view, such as social challenges and psychological challenges. As a result, the challenges facing the desalination industry are most likely six challenges: financial challenges, negative environmental impact challenges, technical complexity challenges, government regulations and permissions, social challenges, and psychological challenges as shown in Figure 5. Each of the above-mentioned challenges will be discussed in the following subsections.

2.1. Financial Challenges

The substantial financial investment necessary for constructing desalination plants is a primary hurdle in their development and operation. A 2008 review of water desalination cost literature identified three factors affecting the cost (in \$/m3) of desalinated water: the type of feed water (seawater or brackish water), the energy source, and the size of the plant, with energy costs exerting the most significant impact on desalination expenses [53].

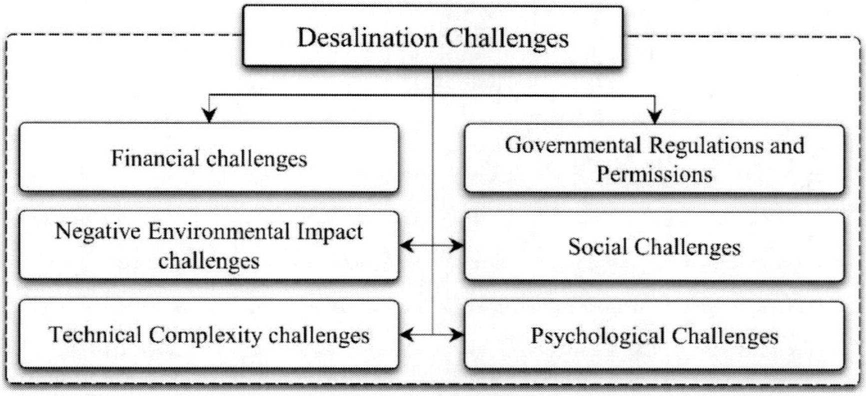

Figure 5. Desalination Challenges.

In 2011, the research projected investment costs for a new desalination facility to range from USD 800 to USD 1500 per unit of capacity (m3/d). Although costs might fluctuate significantly based on local factors such as loan rates, material pricing, and labor expenses, studies estimate the annual average operating and maintenance costs to be around 2–2.5% of the investment [54]. Although there have been considerable decreases in total desalination expenses in recent years, only high- and middle-income nations can finance large-scale desalination technology; it remains prohibitively costly for the majority of low-income countries. At normal conditions, the standard production cost of traditional desalination plants utilizing fossil fuels ranges from USD 1 to 2 per cubic meter, and around USD 0.5 per cubic meter at optimal conditions [55]. Many tender contracts from owners and investors require bidding businesses to provide both CapEx and OpEx in their technological solutions to determine the total production cost per cubic meter of freshwater. Owners and corporations encounter difficulties in validating or comparing the design and actual total production costs of desalination, particularly the OpEx, owing to fluctuations in source water characteristics and plant operating circumstances, whether during startup or after extended periods of operation. The cost of water is calculated by dividing the sum of the amortized capital costs (USD/year) and the annual O&M costs (USD/year) by the total annual desalination plant freshwater production volume

(USD/year). The elements of total water cost, especially for the PPP desalination model, are shown in Figure 6.

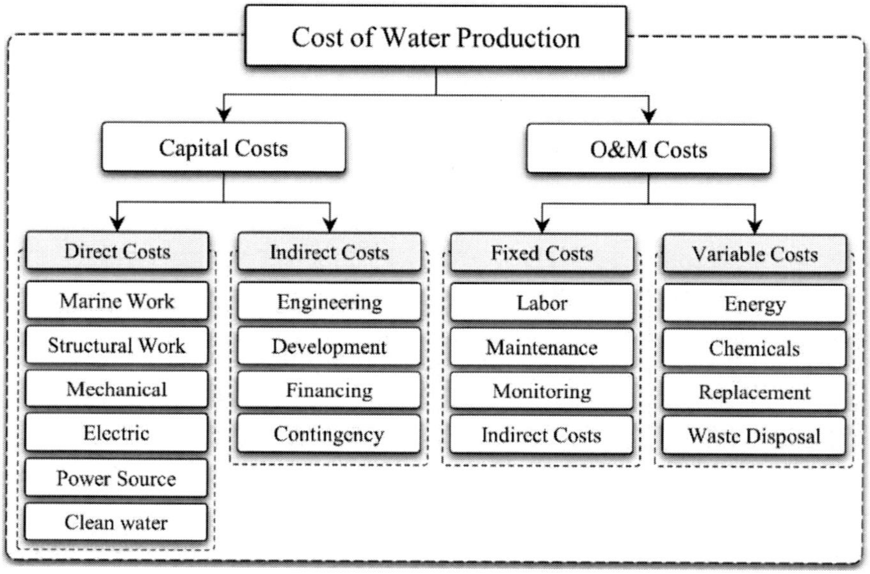

Figure 6. Total Water Cost from Desalination.

For most desalination plants, the CapEx costs and components are divided into nine parts: intake and conveyance of raw water, pretreatment, desalination, post-treatment, pumping, and storing of the product water, electrical and instrumentation systems, buildings, site, and civil works of the plant; the brine discharge and the handling of solids; as well as miscellaneous engineering and development costs. Several other factors must also be considered, such as the type of intake and its location from the plant, the type of intake screens and intake structures, and the type of intake pipelines (buried or above ground). The intake, pretreatment, outfall, and high-pressure pumps commonly used in Reverse Osmosis (RO) desalination use large amounts of energy, making it expensive. Energy costs contribute the most to operational costs for thermal and membrane desalination processes, with almost 40–60%. Multi-Stage Flash (MSF) distillation is not economically competitive for desalination due to the high capital costs. Different materials could gradually accumulate on membranes during operation, organic material such as algae bloom and jellyfish, inorganic solutes, and biofilms contaminating the feed water. The overall effect is a shortage of permeability, which is offset by increasing pressure to maintain water production. Additionally, this counts to the power consumption and water cost over the plant's lifetime. Consumable increase as per inflation and deficiency of equipment. The Desalination Economic Interests (DEI) has different points of view, such as governmental organizations, which are interested in the gap between cost and tariff; investors who are interested in money return; and private sectors, which are interested in keeping investments, as shown in Figure 7.

2.1.1. Machine Failures Affecting Desalination Finances and Economy

Lack of visibility is one of the challenges desalination plants face, which leads to suboptimal timing for maintenance, increased costs, and unnecessary downtime or outages. It is estimated that almost every plant loses at least 5% of its overall productivity, with others experiencing as much as a 20% loss due to downtime (machine failure), downtime is defined as the period when a machine is not in production; it includes any stops during production that cause a loss of revenue for the organization. RO Membrane cleaning is one of the standard reasons causing desalination plant downtime, as it takes the plant offline and causes liquid waste to be discarded. Membrane cleaning is essential to overcome the shortage of permeability, which is offset by increasing pressure to maintain water production. Additionally, this counts to the power consumption and water cost over the plant's lifetime.

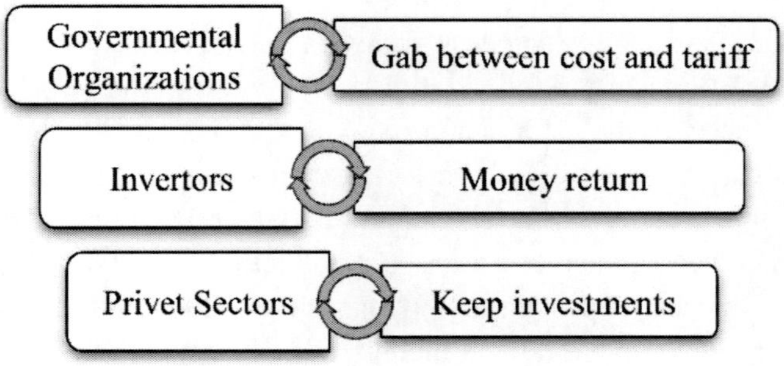

Figure 7. Desalination Economic Interests.

2.2. Negative Environmental Impact Challenges

Desalination is accountable for 0.4% of the global electricity consumption. Although RO is the most energy-efficient seawater desalination technology, it still requires high amounts of energy. Alongside high energy consumption, increased CO_2 emissions are a key challenge of desalination. The estimated RO carbon footprint has been measured from 0.4 to 6.7 kg CO_2 eq/m^3. (1000 m^3 of seawater could potentially release as much as 6.7 tons of CO_2) [56, 57].

Environmental problems may arise during both the building and operation of desalination facilities as outlined below:

- Marine Environmental Pollution: Brine water or Reject water with High Salinity and Brine water or Reject water Chemical Residuals.
- Terrestrial Environment Pollution: Heavy use of chemicals, Cartridge Filter Elements, Membrane Elements, Spare Parts Solids, and Brine treatment technologies.
- Air Environment Pollution: Elevated energy consumption and air pollution resulting from the release of greenhouse gases (GHGs) and air pollutants.

The main three environmental pollution results from desalination plants are shown in Figure 8. Environmental pollution resulting from desalination plants is one of the challenges faced by desalination plants in countries that have unique marine nature, which has an economic impact, especially in the field of tourism, which is considered one of the important sources of national income for such countries.

2.2.1. Marine Environment Pollution

The seawater desalination process removes salt from seawater to obtain freshwater, then typically returns a hypersaline brine to the sea. The salinity of brine produced by the plant is approximately almost double the salinity of ambient seawater. Due to its higher density compared to seawater, brine creates a plume around the outfall and sinks to the seabed unless mixed with ocean water. Hypersalinity can damage or be lethal to marine organisms by interfering with osmotic processes [59], particularly in combination with other stressors, such as temperature and metals [60]. If currents are low, brine can form a persistent layer on the seabed and create a press disturbance to benthic marine communities. Other potential effects of brine on water quality include changes in temperature and dissolved oxygen, although these are generally only issues for thermal plants [61]. Brines can also contain toxic contaminants, such as metals, antiscalants, coagulants and coagulant aids, anti-foaming agents, and cleaning chemicals [62], which can accumulate in sediments around outfalls [63], or in marine organism tissues [64], Additionally, brine can lower pH, which can negatively impact calcifying marine organisms [65]. There are therefore multiple pathways by which brine can impact the marine environment, including changes in salinity, temperature, water quality, hydrology, and pollution with contaminants [66].

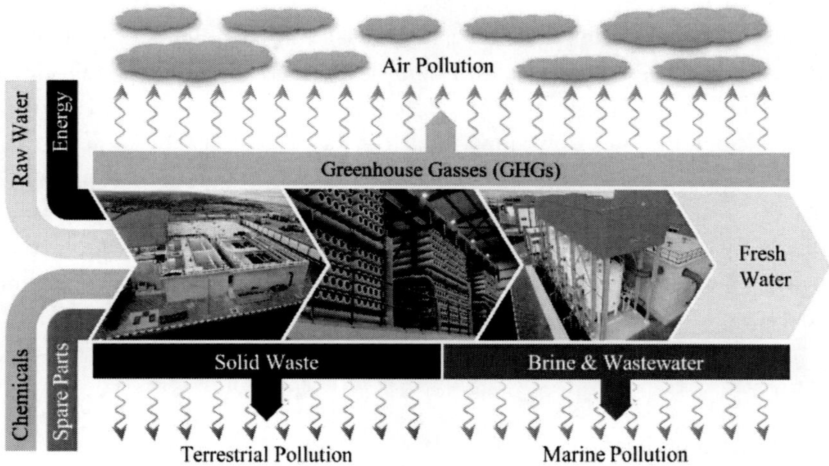

Figure 8. Environment Pollution from Desalination.

2.2.2. Terrestrial Environment Pollution

Desalination is a process that uses heavy chemicals, whether as a primary requirement for the desalination process to produce freshwater to meet the specifications or to treat concentrated water before disposal to sea. Reverse osmosis membranes are also consumed and replaced with new membranes from time to time; cartridge filter elements are used for final purification and replaced at high rates, in addition to spare parts consumed in production processes. All these materials and their residues are disposed of in the ground, so desalination is considered one of the sources of terrestrial environment pollution. This topic requires a lot of research and studies to determine the extent of the negative impact of desalination on terrestrial environmental pollution and how to overcome it.

2.2.3. Air Environment Pollution

Energy is an important aspect of the environmental assessment at desalination plants. Total energy demand includes the energy required for the desalination process, freshwater and brine transportation, plant lighting, office equipment, etc. Since the implementation of desalination on an industrial level, the high energy consumption required has been one of the major obstacles. Despite the recent advancements in desalination technologies, the desalination process is still energy-intensive, and thus the energy consumption should be reduced. To have a better view of the energy consumption, it is interesting to note that the energy needed to obtain 10 L of freshwater through RO (at the energy consumption of 6 kWh/m^3) is equivalent to the power required in a 60-watt incandescent bulb that operates for 1 h. The use of fossil fuels to generate the necessary energy is associated with emissions of GHGs. The thermal-based technologies, namely MSF and Multi Effect Desalination (MED), have at least ten times higher GHGs emissions than RO. Consequently, desalination has significant environmental impacts on air quality. Desalination is an energy-intensive process, and the carbon intensity of electricity generation significantly varies based on the fuel source. Figure 9 illustrates the carbon intensity of electricity generation from various energy sources for reverse osmosis desalination facilities with a capacity of 3,325 Mm^3/day.

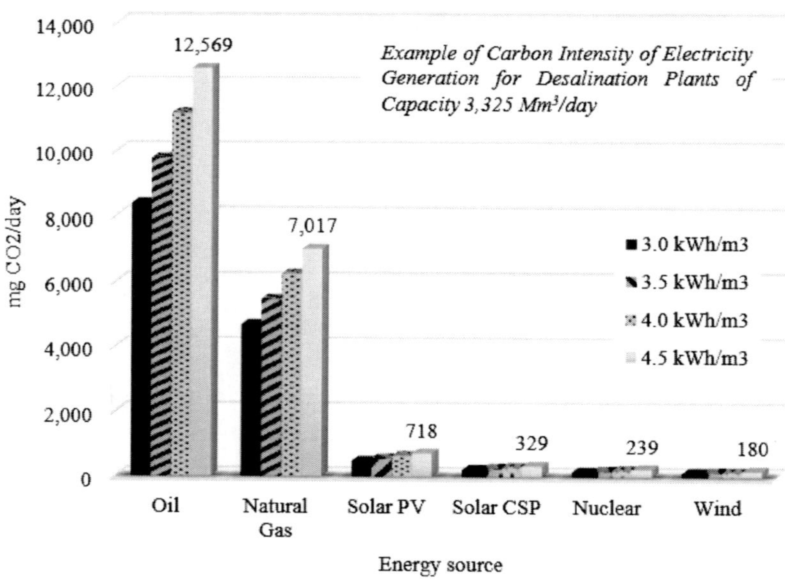

Figure 9. Carbon Intensity of Electricity Generation Using Different Energy Sources.

To make it clearer for readers, carbon emissions for producing one cubic meter of freshwater from desalination using 3.0 kWh/m^3 using oil fuel are equal to driving a patrol car for about 18 km or driving a diesel car for about

15 km, and carbon emissions for producing one cubic meter of freshwater from desalination using 3.5 kWh/m^3 using oil fuel, equal to driving a patrol car for about 21 km or driving a diesel car for about 18 km, carbon emissions for producing one cubic meter of freshwater from desalination using 4.0 kWh/m^3 using oil fuel, equal to driving a patrol car for about 23 km or driving a diesel car for about 20 km and carbon emissions for producing one cubic meter of freshwater from desalination using 4.5 kWh/m^3 using oil fuel, equal to driving a patrol car for about 26 km or driving a diesel car for about 23 km, as shown in Figure 10.

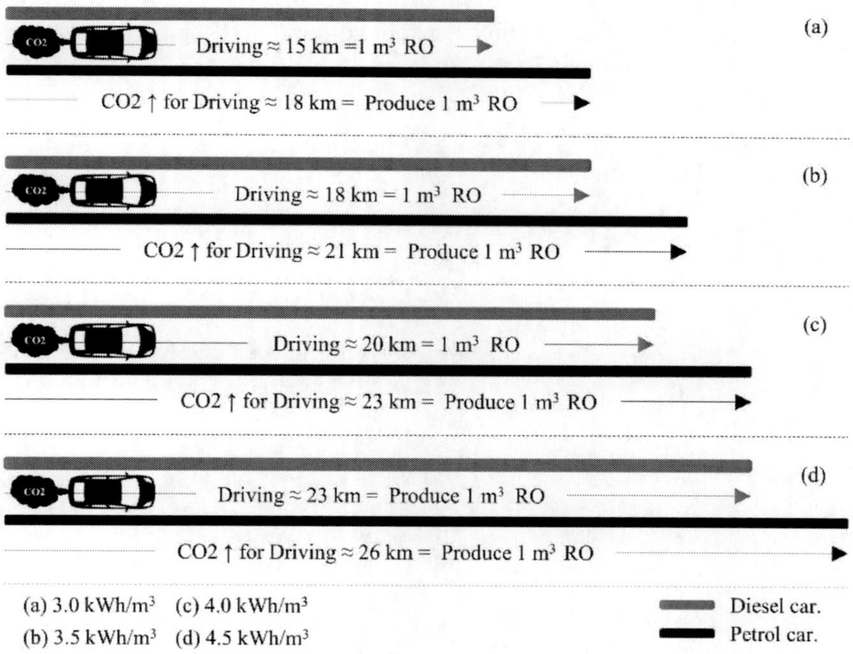

Figure 10. Carbon Emission for Producing One m3 from RO vs Driving Petrol Car or Diesel Car.

2.3. Technical Complexity Challenges

Desalination processes are considered one of the most complex water treatment processes, especially desalination plants that operate with reverse osmosis technology, as the plant contains many stages (intake and outfall system, pretreatment, reverse osmosis, post-treatment, instrumentation, and control, etc.), and these stages include many equipment, pumps, valves, membranes, energy recovery units, instrumentation, etc., and each item of the equipment has its own efficiency and optimal operating conditions. Desalination process is considered a complex process, whether in the design stages or in the operation and maintenance stages, which puts designers under the challenge of designing dozens of equipment in an integrated and compatible system to obtain balanced technical and financial efficiency for the total components while not neglecting the technical efficiency of each element, taking into account the natural changes that occur in the design inputs, it also leaves those in charge of operation and maintenance with the challenge of operating all this equipment in harmony and agreement to produce water with the required specifications while maintaining the efficiency of the equipment in the short and long term. Highly skilled and experienced scientists, engineers, and technicians are essentially needed for this type of process. Figure 11 shows examples of governing equations for some items of desalination plants.

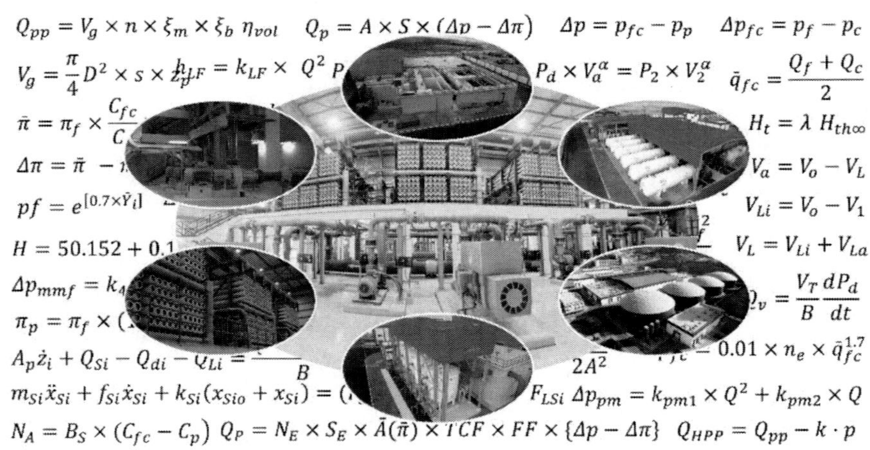

Figure 11. Governing Equations of Some Items of Desalination Plants.

2.4. Governmental Regulations and Permissions

Governmental regulations and permissions major regulatory and permitting issues that relate to planning and implementing desalination plants include developing the saline water source and the discharge and management of concentrate. Federal laws that are pertinent to desalination are briefly described below. The environmental legislation necessitates a reassessment of the regulations governing the desalination business, particularly in light of the swift advancements in desalination technology. Taxation and customs: the necessity to reassess the legislation governing the import and export of equipment pertinent to the desalination sector to promote the development of a domestic desalination industry. The approach of bidding desalination projects requires reevaluation to align with the specific characteristics of desalination plants, hence facilitating project establishment and maximizing advantages.

2.5. Social Challenges

The social challenges of water desalination plants share some points in common and differ in some points from one country to another, according to the nature of the people and their dependence on natural freshwater sources for decades and centuries past. In the case of such countries, water desalination faces the following challenges:

2.5.1. Feeling Unsafe Depending on Desalinated Water

The psychological challenge for citizens to use desalinated water, especially in countries dependent on natural freshwater sources, whose people have depended on water from freshwater rivers and lakes for thousands of years, where peoples feel safe and comfortable natural freshwater sources, for them, the rivers bring safety, water, and nutritional stability, and the sea may bring some pleasure for a short period, but it does not achieve safety, water, or nutritional stability.

2.5.2. Lack of Confidence in Desalinated Water Quality

As a result of some problems and bad experiences with desalination plants, especially in remote areas, such as the small amount of water and the high salinity of desalinated water, and even in cases where the salinity of the water is not high, there is a difference in the taste of desalinated water from the taste of natural freshwater sources, to which people are accustomed. There is also the presence of a color in the desalinated water as a result of poor operations and maintenance in some cases, in addition to the presence of stakeholders from the lack of water, especially in the remote areas, who resist desalination projects in several ways, including spreading rumors about the desalinated water and that it leads to kidney failure and causes infertility.

2.5.3. Feeling Unsafe Following the Public-Private Partnership Model

There are two basic elements regarding this point. The first is the citizens' feeling of insecurity that the private sector owns responsibility for water supply and the fear of the high cost of water in their monthly expenses. The second is the feeling of anxiety about losing government jobs for workers in the field of water desalination.

2.6. Psychological Challenges

2.6.1. Psychological Status of Citizens

The installation of desalination facilities in rural locations experiencing water scarcity and low levels of citizen education has occurred. The water supply in these isolated regions relies on water transportation, vehicular distribution, and the utilization of water bottles. Water distribution in these regions is seen as a revenue source and a favorable position for the individual overseeing the water distribution commerce. Simultaneously, it constitutes an economic burden on the remainder of the population. The beneficiaries in these regions are experiencing the effects of desalination expansion, influenced by the psychological state of the populace due to the dissemination of misinformation regarding desalinated water. The disparity in taste between desalinated water and natural freshwater has led to rumors regarding the detrimental effects of desalinated water, such as "kidney failure or infertility." Additionally, instances of discoloration in desalinated water at certain plants, attributed to technical malfunctions, and operational challenges faced by desalination facilities, along with the cessation of some projects, contribute to these concerns.

2.6.2. Employee Productivity and Safety

People are vital resources for desalination plants. Enabling humans to be as productive as possible is a key concern. In general, human beings are prone to making mistakes, slips, or lapses that often occur when assigned routine tasks that are very typical and require little thought, as example, taking water quality readings from water treatment equipment is a repetitive task that needs to be done periodically by other staff and often leads to many data entry errors. To illustrate, training employees who work with hazardous equipment and risky processes cannot always be manageable. Thus, improving the training quality and employees' productivity while ensuring their safety is a challenge in desalination plants. An important factor causing employee productivity reduction is the absence of immersive training.

2.6.3. Psychological State of the Owner

The preference for foreign product over local products stems from greater trust in there finishing quality and overall form. Certain local enterprises engage in cost-cutting measures to secure deals, irrespective of product quality. Local non-specialist enterprises start operations by producing desalination plants without requisite experience or expertise, resulting in contractual problems with the owner, as desalination projects differ significantly from other water treatment procedures.

3. Desalination Industry Environmental Improvement

Towards Sustainable and Affordable Desalination, some activities and actions have to be taken to overcome the challenges facing desalination. The key areas encompassed in this context are education, training, engineering studies, design optimization, economics, research and development (R&D), localization, social and psychological awareness, and PPP.

3.1. Education

The desalination business needs professional water engineers who are knowledgeable about the specific requirements related to chemical, mechanical, electrical, control, and civil engineering. Consequently, it is imperative for institutions to implement this specialization with the requisite curriculum to train water engineers. These water engineers are experts in processes, installation, operation, and maintenance. This is regarded as an extensive plan; a concise educational strategy involves relying on established university branches with modifications to educational courses across various disciplines, while concurrently, graduation projects will focus on existing water treatment facilities employing diverse methodologies to serve as practical graduation projects. Identical measures must be used in technical education to produce trained technicians through established technical institutions, water experts, and primarily in desalination. This stage seeks to engage competent engineers and technicians [67].

3.2. Training

Desalination training is crucial for acquiring an in-depth comprehension of the mechanisms and scientific principles underlying desalination, hence enhancing the sustainability and cost-effectiveness of current facilities and facilitating the development of innovative technologies. Training must encompass all personnel engaged in desalination, including those involved in policy and planning, procurement of desalination plants, operations and maintenance, research, and academics participating in desalination projects. Training must encompass all proficiency levels, from novices to professionals, and it should remain current with all upgrades. Desalination training provides a comprehensive grasp of the scientific, engineering, and managerial facets of desalination, encompassing the chemistry of water involved in the process and the financial analysis of water costs. Training will enhance knowledge and skill, offering both theoretical and practical insights into maintenance processes and practices. It enhances the operational efficiency of maintenance activities and the psychological and productivity levels of personnel in the desalination sector. Training ensures investment in the desalination sector.

3.3. Engineering Studies

The desalination industry is significantly influenced by engineering studies. Desalination has five primary components: intake and outfall structure, pretreatment, desalination technique, post-treatment, and control system. Each section encompasses a diverse array of techniques; the choice of the appropriate technique varies by project and is contingent upon several factors, including raw water characteristics, desired product water specifications, site conditions, staff expertise, and national regulations. Desalination and water treatment engineering are context-dependent, necessitating engineering studies to identify the optimal desalination procedure. Engineering studies ought to evaluate the existing capabilities inside the country and explore avenues for enhancement.

3.4. Design Optimization

Desalination is among the most expensive and energy-intensive technologies for producing freshwater; optimizing the design of such systems is a complex challenge. Artificial intelligence along with simulation models is utilized to optimize the total cost of producing freshwater per unit. The models account for operational circumstances and technical specifications, as well as accessibility for operation and maintenance, to minimize downtime.

3.5. Economics

Economic analyses for desalination initiatives are crucial to reduce the cost of water production. The nationalization of the desalination business necessitates economic analyses to determine which components should be domestically developed and produced, as well as strategies to reduce costs without compromising performance.

3.6. Research and Development

Research and development (R&D) encompasses two principal models. either assigned to the direct creation of new items or responsible for product development and enhancement. Both models are significant in the desalination sector. Research and development in desalination should primarily concentrate on enhancing the efficiency of current projects, either by lowering production costs or augmenting output with existing infrastructure; promoting local manufacturing of system components and developing innovative technologies or integrated methods to reduce energy expenses, utilizing renewable energy sources such as wind and solar power. Research topics such as desalination cost reduction, integration between renewable and non-renewable energy and desalination process and utilizing artificial intelligence and simulation technologies to optimize the desalination process efficiency and reduce the cost. These research topics should consider the efficiency and sustainability of various desalination methods integrated with renewable and non-renewable sources of energy as an integrated energy source for desalination.

3.7. Localization

Desalination in lower-middle-income economies countries faces the challenge of foreign currency availability and an unstable exchange rate. This challenge exists in both the implementation phase and the operation and maintenance phases. To overcome the currency challenge, it is necessary to localize the desalination industry by encouraging the establishment of local companies to establish, operate, and maintain stations, in conjunction with encouraging the establishment of local factories for the components feeding desalination projects. Encouraging the establishment of local companies results in a 15% reduction in the need for foreign currency. Establishing factories specialized in producing supplies for water desalination plants achieves a saving in foreign currency of 10% of import prices and 25% of manufacturing prices. Localization will lead to: establishing local companies and creating job opportunities outside the country to become a source of national income; introducing foreign currency to the country in addition to using local products in implementing the stations; establishing factories specialized in producing water station supplies to achieve self-sufficiency and export; exporting engineering services; and creating highly trained local workers for self-sufficiency and providing job opportunities outside the country. Considering Egypt's desalination case, localization led to an average reduction in the amount of foreign currency challenges in the period from 2108 to 2021 [68], foreign currency needed amount is as shown in Table 2, while foreign currency saving amounts are as shown in Table 3.

Table 2. Foreign currency needed amount

Item	Implementation	Operation and Maintenance	Total
Foreign Currency Amount Needed (M Euro)	791.5	681.2	1472.2

Table 3. Foreign currency saving amount

Item	Implementation	Operation and Maintenance	Total
Foreign Currency Amount Needed Direct savings (M Euro)	221.4	219.6	441.0
Foreign Currency Amount Needed Indirect savings (M Euro)	included	included	206.0
Total (M Euro)			647.0

3.8. Social and Psychological Awareness

Increase awareness of residents and workers in the field of desalination, to clarify the importance of desalination as one of the elements of securing the water shortage and as one of the sources of economic development, as it has an economic impact on citizens and will improve their economic situation. Raising awareness of the importance of adopting a model of PPP in desalination projects while clarifying that this type of project will lead to the existence of direct and indirect job opportunities by providing jobs in these projects for citizens in the construction areas with lucrative salaries and job opportunities. Indirect activities related to the project, whether in the construction phase or the operation and maintenance phase, in addition to the jobs required for industries and feeding activities such as manufacturing workshops and education and training institutions, will lead to raising the educational, psychological, and economic level for the people. Holding popular awareness seminars and radio and television programs to explain the advantages of these projects, refute rumors related to desalination projects, and clarify the facts.

3.9. Public-Private Partnership (PPP)

As per the economic, social, and development challenges, lower-middle-income economies countries faces, regarding desalination, most of the governments in these countries decided to turn to the model of PPP in desalination projects as one of the basic actions to overcome desalination challenges. From a legal perspective, the governments have to activate or initiate regulating private sector participation in infrastructure projects, services and public facilities, and to form a higher committee for participation affairs and a central unit for participation to overcome the executive obstacles to the law when its issuance or insisted. From an executive perspective, governments have to take several measures, such as cooperating with Sovereign Fund of the local country and international fund firms, allocating land for the establishment of desalination plants, in addition to starting economic feasibility studies for desalination projects including study of using renewable energy as a source of energy.

In this chapter, considering economic and social requirements, a new contracting method under the umbrella of PPP is created, as shown in Table 4. PPP model is expected to improve the desalination industry environment from a socioeconomic point of view, which is considered one of the biggest challenges in desalination industry. The benefits of utilizing the PPP model for desalination are: risk sharing, efficient resource allocation, expertise and innovation, faster project implementation, improved service quality, long-term sustainability, economic growth and job creation, and reduced fiscal pressure. The proposed economic model in Table 4 [68], in terms of involving citizens in the financing process, which is called Public-People-Private Partnership (PPPP) is one of the models that will help overcome the challenges facing the desalination industry, especially the social and psychological challenges.

Table 4. Contracting method under the umbrella of PPP

	Bid Build	Design Build	Design Build Finance	Design Build Finance Operate Maintain	Design Build Multi-Finance Operate Maintain
Identify Infrastructure Need	Public Sector	Public Sector	Public Sector	Public Sector	Public Sector
Propose Solution	Public Sector	Public Sector	Public Sector	Public Sector	Public Sector
Project Design	Public Sector	Privet Sector	Privet Sector	Privet Sector	Privet Sector
Project Financing	Public Sector	Public Sector	Privet Sector	Privet Sector	Public Sector Privet Sector (Sole Proprietors & Companies
Construction	Privet Sector	Privet Sector	Privet Sector	Privet Sector	Privet Sector
Operation and Maintenance	Public Sector	Public Sector	Public Sector	Privet Sector	Privet Sector
Ownership	Public Sector	Public Sector	Public Sector	Public Sector	Public Sector

4. Information Technology and Operational Technology

4.1. Information Technology (IT)

Information Technology (IT) refers to the application of computer systems and devices to process and manage information. IT is a key aspect of the workplace, shaping business processes and individual interaction with information, and it touches on most aspects of daily life. IT is an integral part of our day-to-day existence, involving data storage, data retrieval, access, and data processing. From large corporations to individual entrepreneurs and small-scale businesses, dependency on IT transcends all levels. Multinational companies use it to organize information and make operations more efficient, and even small-time vendors can utilize mobile payment systems for making payments. For example, when we use a spreadsheet to track the purchasing of gifts, we are indeed using information technology.

The phrase "Information Technology" refers to many types of technological methods of communication, data transmission, and information handling. Trends associated with it are many, which include analytics, automation, artificial intelligence, cloud computing, communication systems, cybersecurity, data and database administration, infrastructure, the Internet of Things (IoT), machine learning, maintenance and repair services, networks, robotics, software development, applications development, and Supervisory control and data acquisition (SCADA).

IT supports most working and personal activities, the foundation for communication, technological progress, innovation, sustainability, and entertainment. At a personal level, it supports communication, gaming, sharing media, online purchases, and social interaction. At a business level, IT is used in the running of businesses in almost all industries from food service and healthcare to retail and manufacturing. Additionally, in areas such as desalination, information technology plays a fundamental role in the handling of information and process improvement [69-72].

4.2. Operational Technology (OT)

Operational Technology (OT) is hardware and software that detects or causes a change, through the direct monitoring and/or control of industrial equipment, assets, processes, and events. OT is the hardware and software that monitors and controls devices, processes, and infrastructure and is used in industrial settings. IT combines technologies for networking, information processing, enterprise data centers, and cloud systems. OT devices control the physical world, while IT systems manage data and applications. OT is for connecting, monitoring, managing, and securing an organization's industrial operations.

Businesses engaged in activities such as manufacturing, water desalination, mining, oil and gas, utilities, and transportation, among many others, rely heavily on OT. Robots, industrial control systems (ICS), SCADA systems, programmable logic controllers (PLCs), and computer numerical control (CNC) are examples of OT. The key difference between IT and OT is that IT is centered on an organization's front-end informational activities, while OT is focused on its back-end production machines.

OT refers to hardware and software that is used to monitor and control industrial processes, equipment, and events, and make changes through direct control. OT plays an essential role in industry for managing devices, infrastructure, and processes. IT refers to technology that is applied in networking, data processing, enterprise data centers, and cloud computing. While OT gear addresses the physical aspects, IT solutions focus on managing data and applications. OT has to connect, operate, and safeguard industrial operations in organizations. Industrial sectors that heavily rely on OT include manufacturing, water desalination, mining, oil and gas, utilities, and transport. Some examples of OT are robots, industrial control systems ICS, SCADA systems, PLCs, and CNC systems. The major distinction between IT and OT is their concern: IT deals with front-end information operations, while OT deals with back-end production operations.

5. Integration of Information Technology and Operational Technology for Desalination

For decades, to safeguard OT equipment, most operational and maintenance companies maintained a physical separation from external networks, referred to as an air gap. Nevertheless, the growing advantages of interconnected operational technology gadgets are prompting several firms to abandon that strategy. An integrated IT/OT system may optimize desalination production operations and enhance sustainability; however, these advantages also introduce heightened dangers.

Organizations confront challenges in merging IT and OT sectors due to the lack of unified software interfaces, defined data formats, and communication protocols. The discrepancies have caused the IT and OT sectors to seem like separate universes.

Desalination systems comprise a substantial array of instruments and equipment. A seawater desalination facility with a capacity of 40,000 m^3/day comprises 329 machines, 118 instruments, 146 control valves, and 64 pieces of equipment. It comprises 928 input/output points, consisting of 565 digital input points, 182 digital output points, 153 analog input points, and 28 analog output points, as seen in Figure 12.

This equipment and these devices encompass a wide range of data, variables, and factors affected by operating parameters, equipment performance, and water production goals.

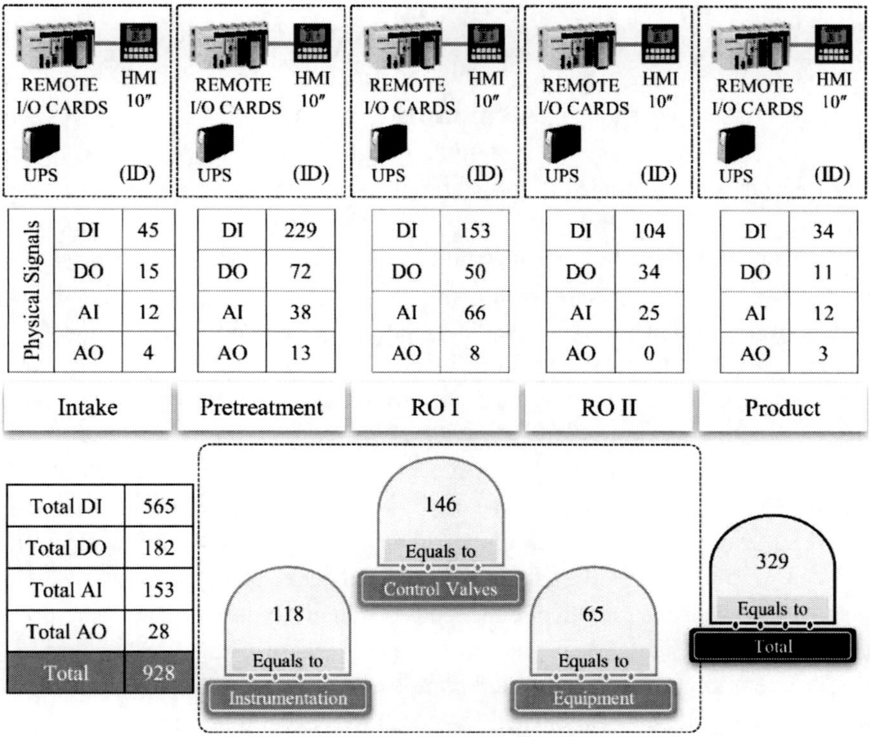

Figure 12. Machines and Input / Output Points for 40,000 m^3/day SWRO System.

Expert operators, engineers, and administrators are essential for overseeing daily data and fluctuations. The examination of this complex data can be challenging, preventing plant managers and engineers from swiftly discovering appropriate solutions. Machine Learning (ML), a branch of Artificial Intelligence (AI), can evaluate vast production datasets to forecast probable results and adjust accordingly.

This can convert problems into solutions, improve circumstances, and augment efficiency. By cultivating a suitable attitude and mindset, challenges in the desalination industry may be converted into opportunities; nevertheless, it is crucial to first evaluate them from a larger perspective. These issues may create a rift between IT and OT. The disparity between the two technologies compels desalination firms to contend with IT and OT until a unified software interface can be established. Progress in Internet of Things (IoT) technology and AI enables the digital information realm to perceive, comprehend, and impact the physical operating environment. When executed effectively, IT and OT integration may consolidate business operations and insights from a production setting into a cohesive ecosystem.

6. Smart vs Intelligent Desalination Systems

Smart desalination denotes the application of advanced technology, including sensors, automation, and data analytics, to enhance the operation of desalination facilities. The primary aims are enhanced efficiency, decreased energy use, and superior quality of the final water production. It is implementing sophisticated control systems associated with Industry 4.0, which focuses on the integration of IT and OT systems. The technology and software comprising these systems can autonomously identify defects and notify operators of potential issues. The IoT is enabling the creation of sophisticated gadgets equipped with clever sensors that continuously transmit data to the cloud for comprehensive analysis. Data are examined via ML and AI, which gain potency as additional data are amassed. Water production facilities employing intelligent, interconnected technology and data-driven decision-making can achieve enhanced productivity and efficiency. IoT sensors deliver instantaneous data on several facets of water flow, including pressure, temperature, and chemical composition. They monitor the status of equipment by the assessment of temperature, vibrations, and operational parameters. Various components, including pipes, pumps, and valves, may be integrated with IoT to facilitate seamless communication during the desalination process. While IoT focuses on collecting real-time data, AI and ML employ mathematical algorithms to derive insights from that data. The substantial amount, velocity, and diversity of data generally necessitate AI-driven big data analytics solutions to enhance industrial efficiency and decision-making.

Smart desalination, conversely, is a sophisticated system that employs artificial intelligence, machine learning, and deep learning to create a self-aware, self-sufficient, and adaptable mechanism. Intelligent desalination involves predictive analytics using AI, real-time optimization, automated decision-making, and ongoing learning. It is based on advanced technologies like AI, ML, deep learning (DL), and natural language processing (NLP), providing optimal performance, predictive maintenance, autonomous operation, and ongoing enhancements. Smart desalination and intelligent desalination are markedly different. Intelligent desalination exhibits more autonomy by utilizing artificial intelligence for control and decision-making processes. Intelligent desalination heavily depends on artificial intelligence and machine learning, whereas smart desalination focuses on automation and data analysis. Moreover, the optimization scope in intelligent desalination encompasses the whole desalination process, whereas smart desalination focuses on optimizing certain parameters, such as energy consumption or water quality. Intelligent desalination optimizes some aspects of the desalination plant process, representing a progression in the operational evolution of desalination facilities. Intelligent desalination is a transformative advancement in the development of an autonomous, self-monitoring, and self-adaptive desalination system.

Conclusion

Desalination is the most feasible and promising solution to provide freshwater, especially for countries that haven't enough freshwater resources or have a shortage of freshwater resources. There are four main challenges facing the water desalination industry: financial challenges, negative environmental impact challenges, technical complexity challenges, and government regulations and permissions.

Countries that depend on natural water sources such as freshwater rivers and lakes to provide water to citizens, meanwhile have lower-middle-income economies facing more challenges affecting the desalination industry, which are: social challenges: feeling unsafe depending on desalinated water; lack of confidence in desalinated water quality; feeling unsafe following the public-private partnership model; and psychological challenges: the psychological status of citizens; employee productivity and safety; and the psychological state of the owner.

The PPP model has been considered a robust and effective mechanism for overcoming desalination challenges in countries that have lower-middle-income economies, supporting economic growth, infrastructure development, and service delivery across the desalination sector. Some key benefits of utilizing the PPP model, especially for desalination, are: risk sharing, efficient resource allocation, expertise and innovation, faster project implementation, improved service quality, long-term sustainability, economic growth and job creation, and reduced fiscal pressure.

Activities and actions have to be taken to achieve sustainable and affordable desalination by both the governments and private sectors to overcome the challenges facing desalination, such as education, training, engineering studies, design optimization, economics, research and development, localization, social and psychological awareness, and the PPP financial model. From a legal perspective to utilizing the PPP model, the governments have to activate or initiate regulating private sector participation in infrastructure projects, services, and public facilities, and form a higher committee for participation affairs and a central unit for participation to overcome the executive obstacles to the law when its issuance or initiated.

From an executive perspective to utilizing the PPP model, governments have to take several measures, such as cooperating with the Sovereign Fund of the local country and international fund firms, allocating land for the establishment of desalination plants, in addition to starting economic feasibility studies for desalination projects with a study of using renewable energy as a source of energy.

The proposed economic model of Public-People-Private Partnership (PPPP), in terms of involving citizens in the financing process, is one of the models that will help overcome the challenges facing the desalination industry, especially the social and psychological challenges.

Chapter 3

Desalination Systems Plant Performance Index

Abstract

Recent advances in seawater desalination methods have added to the feasibility of desalination as a municipal water supply source. Over the decade since 2010, the number of desalination plants worldwide has grown by 4.8% annually, adding close to 4.8 million m^3/day to production capacity. 20,971 projects as of mid-February 2020 had an installed capacity of 97.2 million m^3/day, which increased to 115.01 million m^3/day in 2024. High costs associated with CapEx and OpEx remain significant challenges for the desalination industry, impacting the price of produced water. Although demand for desalinated water often rises with water shortages, price becomes the primary driver of demand as these needs diminish. Many factors make cost estimation complex and site-specific, including environmental constraints, energy costs, and plant technology. The Performance Index for Reverse Osmosis Desalination Plants (RO_{PI}) is a metric designed to evaluate how effectively RO plants achieve their water production goals, aims to provide a standardized measure of performance across all plant sections by assessing the Actual Cost (AC) to Planned Cost (PC) ratio for producing one cubic meter of freshwater. This index will help stakeholders determine plant performance concerning technical efficiency and production costs, taking into account other factors such as feed water quality and maintenance practices.

Keywords: desalination, reverse osmosis, performance index, desalination performance index

1. Introduction

Recent strides in economically viable seawater desalination methods have significantly improved desalination process's feasibility as a potential contributor to municipal water supply. Accordingly, the global growth in the number of desalination plants has been phenomenal, recording an average annual growth rate of about 6.8% since 2010, which translates into an annual increase of approximately 4.6 million m^3/day in production capacity. In fact, between January 2019 and February 2020, contracts were awarded, and installations of 155 new desalination facilities were installed worldwide, adding another 5.2 million cubic meters a day to the cumulative capacity [73]. The worldwide cumulative capacity for producing freshwater by desalination as of July 2016 was 95.6 million m^3/day, facilitated by 18,983 plants and projects. This cumulative capacity is derived from both existing installation capacity and contracted projects.

As of the end of 2017, considering facilities that have been operational since 1965, the global cumulative capacity for desalination had attained 99.8 million m^3/day, with the effective capacity of the functioning plants projected to represent around 93% of the overall installed capacity. Also, as of the middle of February 2020, the total cumulative capacities for producing freshwater were 114.9 million m^3/day for 20,971 projects [49]. According to International Desalination and Reuse Association (IDRA), the Global cumulative contracted capacity by the year 2024 is 115.01 million m^3/day.

Unfortunately, building and operating desalination plants is expensive, which raises the cost of the water produced [74]. With major progress in membrane desalination technologies, these costs could go down by a lot for both capital and running these plants. The economic feasibility of desalination technology mainly depends on the lifespan of the systems and their capital and operating expenses. The demand for desalinated water is strongly related to the final cost of the produced water. In regions where there is a water shortage, the economic need often drives the demand for desalinated water, regardless of its cost. However, as the economic need decreases, it is the water's price that mostly determines the demand for desalination [46]. Moreover, the high uncertainty cost estimation for both the construction and operation of desalination plants, together with the erratic changes in weather patterns, makes it challenging to predict whether a newly commissioned desalination facility will be viable or inadequate in the long run [44]. Additionally, the costs of desalinated water are subject to significant fluctuations. Because each country that runs these plants has its own unique geography and economy, their overall efficiency could be greatly increased [75].

There are also important political and environmental factors that affect the final cost of desalination, as well as the type of project contract, the cost of labor and energy, the plant's capacity, the amount of salt in the feed water, the method of desalination chosen, and some other smaller factors. That shows that desalination costs are site-specific, so there is no single way to estimate them. Hence, the selection of the most appropriate technology for any location needs to be done systematically. The challenge is that the methodologies developed so far lack precision and are applicable under specific conditions because they cannot embrace all the factors affecting them.

Because of the huge financial requirements for infrastructure projects, many governmental authorities attempt to find alternative funding arrangements. Traditionally, financing capital expenditure by governments has been largely done through the issuance of debt, which is subsequently repaid using revenue generated from tariffs and water charges [76, 77].

This type of debt financing allows the costs to be spread over a period that more closely matches the useful life of the asset, usually twenty to thirty years, and in some cases, even longer. There has been a lot of work done to find new ways to manage traditional funding, though, because construction costs are going up and it's getting harder for government agencies to get money [78, 79].

With the continuous growth in the global production of water through desalination plants, it is urgent to find a single standard to measure how well plants that use this technology are working. Everyone involved in the desalination industry must follow this standard, including companies that make plant parts, companies that offer solutions, people who are in charge of running and maintaining plants, owners, and investors.

In this chapter, we propose to have a Performance Index for reverse osmosis desalination plants RO_{PI}. The suggested performance index RO_{PI} tool is a concept of estimating the production cost to its defined cost value along the plant's lifetime.

2. General Background

The capital cost of a reverse osmosis (RO) desalination plant and its annual operating costs can be considered the two major components of a financial model. Operating costs, often referred to as OpEx, are mostly dominated by energy costs associated with the operation of the desalination plant and are subject to fluctuations in energy costs. Other than energy costs, OpEx encompasses a variety of costs, including labor costs, spare parts, chemical supplies, membrane replacements, and insurance costs [77, 78].

On the other hand, CapEx comprises direct and indirect financial outlays. Direct capital expenses include costs incurred to purchase major machinery, such as high-pressure pumps and other related components, land acquisition, and engineering. The indirect capital expenses include, among others, transportation and insurance and construction and overhead expenses [79].

To find the total water cost (TWC) of desalination at a certain plant, add up the plant's CapEx, which is reduced over its useful life, and its annual OpEx. The resulting total is then divided by the average annual volume of desalinated water produced by the plant [80, 81].

From most expensive to least expensive, the main things that affect operating and maintenance (O&M) costs are technical problems with RO desalination plant systems, management and financial problems, the general features and layout of desalination plants, and government fees. The most important factors were those of feed water characteristics during the pretreatment stage for the different intakes of water. These factors are significant because they affect the O&M costs of desalination plants. This is especially important because it changes according to the location of the plant and needs deep study. These would be followed by the rates of replacement of the membranes, design-related problems, a lack of adequately trained maintenance personnel, and energy consumption among other contributors.

From an overall perspective, both aspects of technical and human resources are presented. The performance index is a conceptual measure of how well work is done concerning its goals. In the early stages of the widespread use of RO technology, many studies were performed to establish methods of determining the operating effectiveness of RO plants. Most of the methods focus on how well the membrane can produce freshwater, going into more detail about the amount of water produced as well as quality factors like membrane flux and salt rejection rates. These parameters turn out to be quite important to the success of the RO unit. In addition, the energy and chemicals used during the manufacture of one cubic meter of freshwater are the other main factors that provide evidence of how effective the system of RO is [82]. Desalination plants utilizing RO technology are typically comprised of three principal components: pretreatment, RO, and post-treatment. Each component encompasses various items that are primarily determined by the characteristics of the water source and the specifications required for the treated water. It is important to consider the water intake as a critical element within the pretreatment phase, irrespective of the specific type of intake utilized. The performance of each part in the RO desalination plant has a big effect on how well and how efficiently the whole system works [82].

The performance index is a suggested tool that evaluates exactly how a work is meeting its planned goal. For a long time, much research was focused on figuring out a way to measure desalination plants' performance as well as the planned and actual cost. Most of the methods worked on the assessment of the membrane's performance to produce high-quality potable water, taking into consideration the membrane's performance in terms of flux and salt rejection, since the membrane is the core of the RO unit. The energy and chemicals required to produce one cubic meter of freshwater are then counted as a key indicator of the RO system's efficiency.

RO Desalination plants generally involve three main stages, Pretreatment, Reverse Osmosis, and Post post-treatment. Each stage is designed mainly based on the feed water characteristics & the specifications of the water needed from the treatment process. In this chapter, we consider the water intake as an item in the pretreatment stage, regardless of intake type.

3. Objectives

It has become common for owners and stakeholders, over the last two decades, to ask bidding companies not only for CapEx -related information but also OpEx analyses related to plant operation. There are cases where CapEx is lower than OpEx and vice versa in other projects, As shown in Figure 13. Lastly, it is necessary for the total production cost of freshwater per cubic meter, including CapEx and OpEx, to be calculated to make informed decisions.

Validating or comparing the design and actual total production costs of RO systems is difficult for owners and companies because source water characteristics such as salinity, temperature, and TSS, in addition to operational conditions, can change during the start-up period of a plant and over its operational life. There needs to be a consistent way to evaluate the performance of plants in integrated systems like RO, which have many different pieces of equipment and parts, like high-pressure pumps, blowers, and filters, each with its own set of performance indicators.

The challenge now is to come up with one single performance index that encapsulates the performance of the whole system with a dynamic system instead of the usually used static one. Hence, RO_{PI} should be developed to estimate the performance of the three main parts of the RO desalination plants using different kinds of equipment with their own performance criteria. Over the past two decades, a majority of owners and investors have submitted documents requesting the bidding companies to include both plant CapEx and OpEx in their technical solutions. This is done to calculate the total production cost of freshwater cubic meters (CapEx & OpEx). This challenge arises from variations in feed water parameters and plant operating conditions, which can be evaluated at the early stage of the plant operation or any point after several years of operation. There is a need to develop a guideline that will enable plant owners and also companies to evaluate their effectiveness of operation, especially in a complex integrated system such as RO system, which consists of several units of equipment, each having its own set of performance measures.

The complexity is a major drawback in the formulation of a comprehensive index that represents the performance of the overall system. In this chapter, we try to formulate RO_{PI} that will evaluate the efficiency of the three main components of the RO desalination plant. Since technical performance and cost are inversely related in the sense that lower performance would generally mean higher costs and vice versa, we will focus on comparing planned vs. actual production costs per cubic meter for the entire RO system.

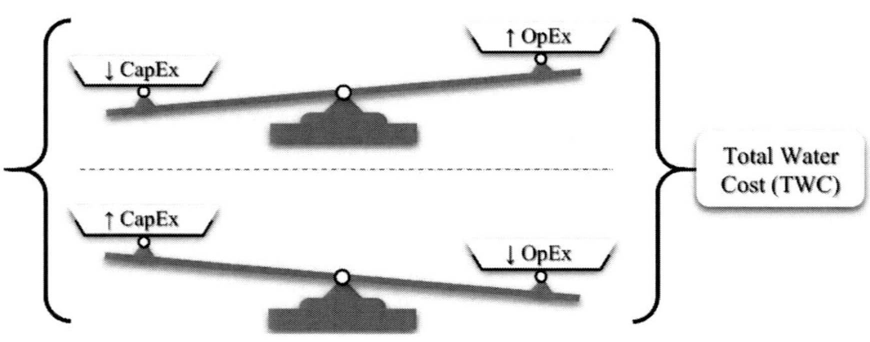

Figure 13. CapEx and OpEx Cases for RO System.

4. Methodology

The performance is based on specified production cost goals over the operating plant life. A new performance index RO_{PI} is defined based on the actual cost (AC) to planned cost (PC) ratio to produce one cubic meter by the whole RO system.

The cost of water production in an RO desalination facility is influenced by many fluctuating factors related to specific designs and locations. Among these are feed water quality, including salinity, turbidity, pH, COD, BOD, etc.; feed water temperature; membrane degradation; chemicals needed;

maintenance and operation schedules; downtime periods; and the skill levels of operators.

The value of PC may be calculated by following two methods, either from the design or startup parameters. PC, calculated based on design parameters, typically varies due to variations in feed water and temperature, as well as membrane aging, resulting in different values. This is why the method gives you two choices: you can choose the minimum, maximum, or mean value, or you can get a time-varying function, PC(t), that shows the planned cost as a function of time. This makes RO_{PI} change over the life of the plant, as shown in Figure 14.

On the other hand, PC calculated from startup parameters makes use of real-time data on feed water parameters, membrane lifetime, and feed water characterization at any given period. Since the fluctuations of the parameters in real operations are unpredictable and affect the costs directly, a compromise between the two approaches is suggested as follows:

- Construct a cost curve to represent production costs over the lifetime of the plant as a function of design variables, denoted PCD(t).
- Construct the cost curve, modeling the production costs through the whole operational life of the plant by taking all start-up parameters but adapting to the trend of the design parameter graph and referring to it as PCS(t), which shall be regarded as a referential basis of the subsequent evaluations.

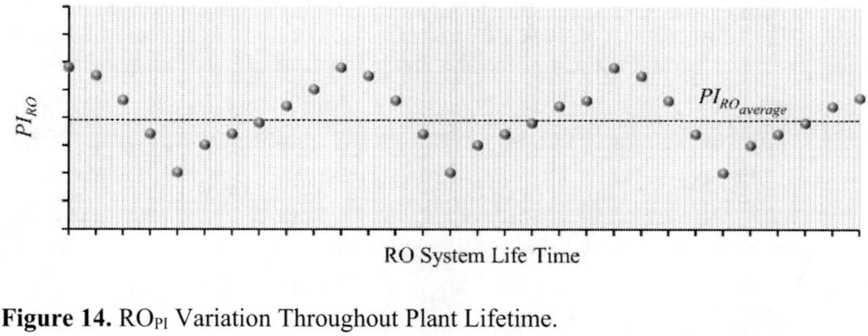

Figure 14. RO_{PI} Variation Throughout Plant Lifetime.

Similarly, we can define a Design Performance Index (RO_{DPI}) for the RO system to evaluate its design effectiveness. All in all, the ultimate performance of the system would lie in the proper and effective design and functional integrality of each subsystem and component. The RO plant needs to efficiently produce the required volume of drinking water while sustainably maintaining acceptable quality levels with minimal cost [2].

5. Case Study

This chapter examines a desalination plant with a seawater treatment capacity of 20,000 m³/day. The plant is made to work as a whole. It has an open intake, normal pretreatment steps, RO boosted with an isobaric energy recovery system, post-treatment with an up-flow calcite contactor, and a product pumping station at the end. The Engineering, Procurement, and Construction (EPC) costs associated with the construction of this plant are estimated to be approximately 26.13 million USD.

The costs of using RO to make potable water from brine sources are also looked at. These costs include both capital expenditures and operating expenses, as well as a lot of different parameters and factors that affect the process.

5.1. Capital Expenditure (CapEx)

In this chapter, the plant's CapEx is broken down into three main parts: structures for inflow and outflow (no matter what kind they are), electromechanical systems, and civil works, which include storage tanks.

The capital expenditure earmarked for intake and outfall structures is about 7.03 million USD, representing 26.92% of the total EPC costs. On the other hand, electromechanical works cost around 16.03 million USD, equivalent to 50.00% of the total EPC cost. In addition, civil works, including all the storage tanks, amount to 6.03 million USD, totaling 23.08% of the total EPC cost. The breakdown of the expenditure in the three categories is depicted in Figure 15.

Further, this chapter, in analyzing the total production costs, acknowledges the uncertainty in the expected lives of the various components. In fact, it is assumed that the intake and outfall structures will last for fifty years, the electromechanical processes are expected to last twenty years, and thirty years for civil engineering works.

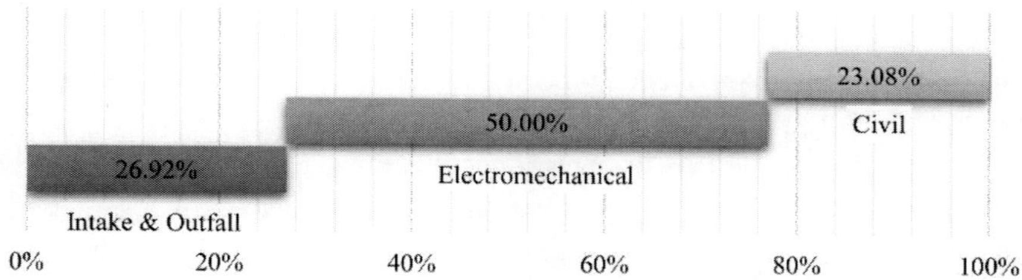

Figure 15. Expenditure Percentage for the Three Main Sections in RO System.

5.2. Operating Expenses (OpEx)

The OpEx analysis in the current chapter includes several plant operating expense components, generically termed O&M expenses. The costs included are the nine major categories listed: intake and outfall O&M, facilities maintenance, process equipment O&M (including spares), power consumption, chemical consumption, membrane replacement, cartridge replacement, labor, and miscellaneous.

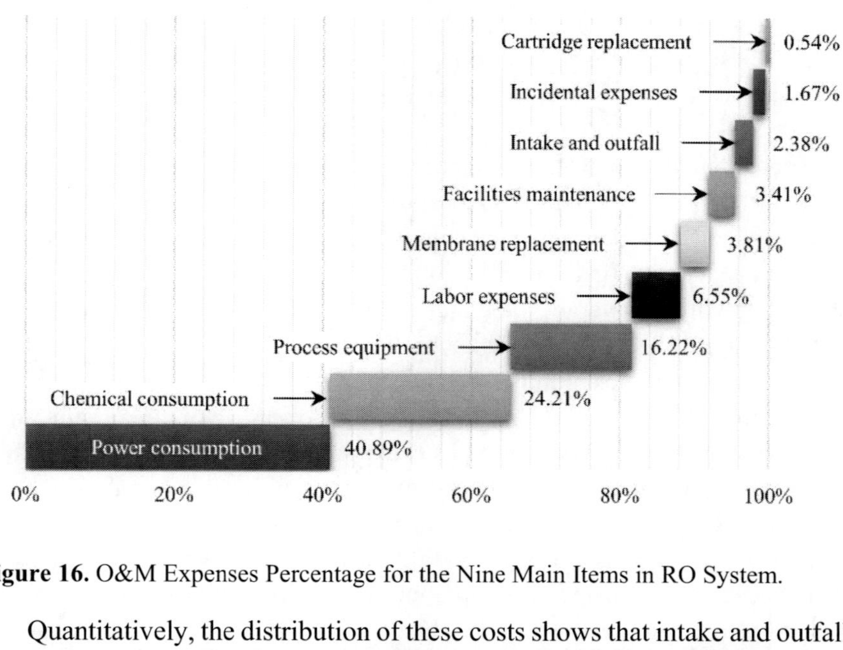

Figure 16. O&M Expenses Percentage for the Nine Main Items in RO System.

Quantitatively, the distribution of these costs shows that intake and outfall O&M costs around 0.009 USD/m^3, which represents 2.38% of the total OpEx.

Facilities maintenance costs are estimated at 0.013 USD/m³, representing 3.41% of the overall OpEx. The costs related to process equipment (including spare parts) are approximately 0.062 USD/m³, representing 16.55% of the total OpEx.

Power consumption is a major cost, estimated at 0.153 USD/m³, representing 40.89% of the total OpEx. In addition, chemical consumption costs are estimated at about 0.091 USD/m³, representing 24.21% of the total OpEx. Membrane replacement has costs of about 0.014 USD/m³, representing 3.81% of the total OpEx, while cartridge replacement costs are negligible at about 0.002 USD/m³, representing 0.54% of the total OpEx. Labor costs are estimated at 0.025 USD/m³, which accounts for 6.55% of the total OpEx, and miscellaneous costs are estimated at 0.006 USD/m³, representing 1.67% of the total OpEx. The breakdown of these cost percentages for the nine major items is shown in Figure 16.

6. Reverse Osmosis System Performance Index

The Performance Index is a metric to assess the quality of operational performance compared to set objectives. RO desalination plants have four main components: intake, pretreatment, reverse osmosis, and post-treatment. Each of these components has some pieces of equipment and processes, and these depend on the quality of the water source and the specifications to which the treated water is needed.

There should be a systematic approach for both operators and owners of plants to assess the overall performance of their plants in the overall integrated system of the RO system, given the complexity of the system and the multitude of components with each having its own set of performance indicators. Coming up with an overall index that best describes the performance of the overall system is a significant challenge.

In order to treat this problem, one must develop RO_{PI}, which characterizes the operating efficiency of all four primary components of an RO desalination plant. Each section is influenced by its corresponding device, which may have variations in performance. The performance and operational cost are interrelated in that a loss in performance generally corresponds to a rise in cost, and vice versa. This interrelationship necessitates comparative examination of anticipated versus real production cost of one cubic meter of product from the entire RO system. Hence, we suggest that the RO_{PI} consists of four basic elements, as illustrated in Figure 17.

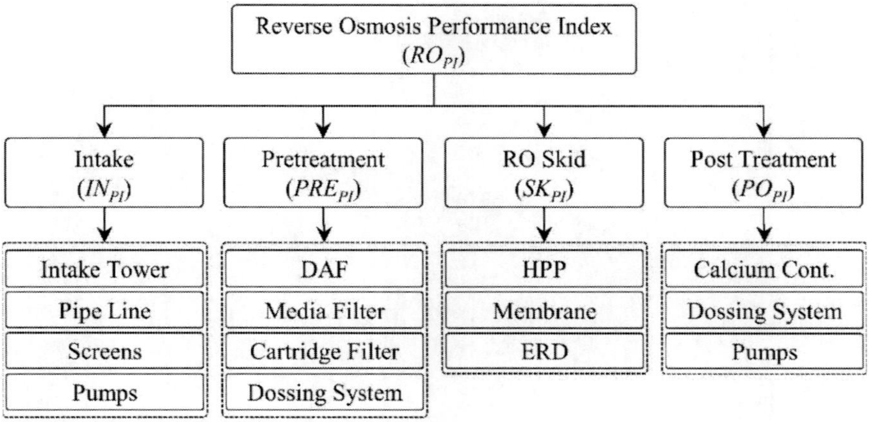

Figure 17. Reverse Osmosis Performance Index Main Items.

7. Performance Index Calculation

The Performance Index value is a function of the Actual Total Water Cost (*ATWC*) and the Planned Cost (*PTWC*) for one cubic meter produced by the entire RO plant. The Planned Total Water Cost, also known as the RO plant Planned Total Water Cost (RO_{PTWC}), includes Planned Capital Expenditures (RO_{PCapEx}) and Planned Operational Expenditures (RO_{POpEx}). By contrast, the Actual Total Water Cost, here referred to as RO plant Actual Total Water Cost (RO_{ATWC}), considers both actual capital expenses (RO_{ACapEx}) and actual operating expenses (RO_{AOpEx}), as shown in Figure 18.

The suggested RO_{PI} is shown below. It measures the difference between the RO plant's RO_{ATWC} and its RO_{PTWC}. The suggested simple example for performance index calculation represents a desalination plant that consists of four main sections. Intake section: include intake tower, pipeline, screens, and intake pumps. Pretreatment section: includes DAF, media filter, cartridge filter, and pretreatment dosing system.

The RO skid section includes a high-pressure pump, a membrane, and an energy recovery device. Post-treatment sections include a calcium contactor, a post-treatment dosing system, and product pumps.

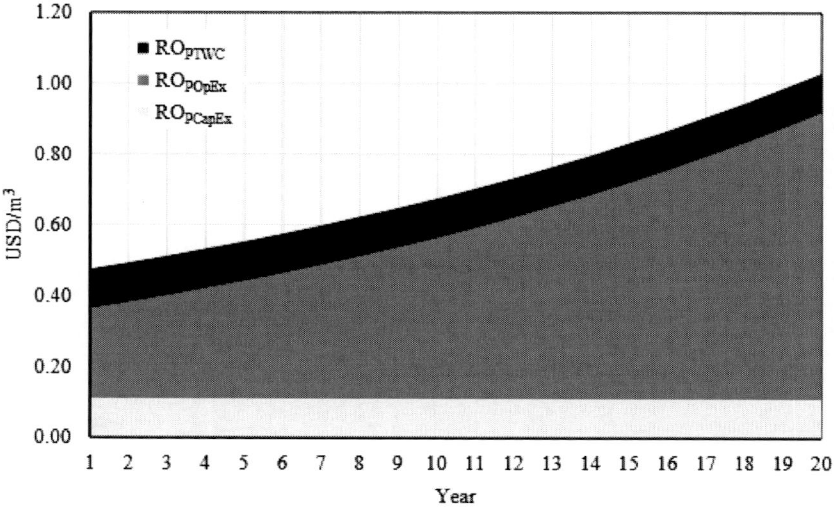

Figure 18. RO plant Planned Total Water Cost Include Both CapEx and OpEx.

The performance index calculation concept is based on the calculation of each item's performance index individually based on cost reference and sum items per section. The proposed Performance Index for RO_{PI} consists of four main sections, as shown in Figure 17. represent Performance Index four main sections are represented by Equation (1).

$$RO_{PI} = IN_{PI} + PRE_{PI} + SK_{PI} + PO_{PI} \quad (1)$$

where: RO_{PI}: Reverse osmosis performance index, IN_{PI}: Intake performance index, PRE_{PI}: Pretreatment performance index, SK_{PI}: RO skid performance index and PO_{PI}: Post-treatment performance index.

RO system performance index based on the comparison between planned total water cost and actual water cost are represented by Equation (2).

$$RO_{PI} = \frac{RO_{ATWC}}{RO_{PTWC}} = \frac{IN_{AWC}}{IN_{PWC}} + \frac{PRE_{AWC}}{PRE_{PWC}} + \frac{SK_{AWC}}{SK_{PWC}} + \frac{PO_{AWC}}{PO_{PWC}} \quad (2)$$

where: RO_{PI}: Reverse osmosis performance index, RO_{ATWC}: Reverse osmosis actual total water cost, RO_{PTWC}: Reverse osmosis planed total water cost, IN_{AWC}: Intake actual water cost, IN_{PWC}: Intake planed water cost, PRE_{AWC}: Pretreatment actual water cost, PRE_{PWC}: Pretreatment planed water cost,

SK_{AWC}: RO skid actual water cost, SK_{PWC}: RO skid planed water cost, PO_{AWC}: Post-treatment actual water cost, PO_{PWC}: Post-treatment planed water cost.

Intake performance index (IN_{PI}), considering intake tower, pipe line, screens and intake pumps. The presented equations for IN_{PI}, are as shown in Equation (3).

$$IN_{PI} = IT_{PI} + PL_{PI} + SC_{PI} + IP_{PI} \quad (3)$$

where: IN_{PI}: Intake performance index, IT_{PI}: Intake tower performance index, PL_{PI}: Pipe line performance index, SC_{PI}: Screens performance index and IP_{PI}: Intake pumps performance index.

The presented equations for Intake tower performance index (IT_{PI}) are as shown in Equation (4).

$$IT_{PI} = \frac{IT_{AWC}}{IT_{PWC}} \quad (4)$$

where: IT_{PI}: Intake tower performance index, IT_{AWC}: Intake tower actual water cost and IT_{PWC}: Intake tower planed water cost.

The presented equations for Pipe line performance index (PL_{PI}) are as shown in Equation (5).

$$PL_{PI} = \frac{PL_{AWC}}{PL_{PWC}} \quad (5)$$

where: PL_{PI}: Pipe line performance index, PL_{AWC}: Pipe line actual water cost and PL_{PWC}: Pipe line planed water cost.

The presented equations for Screens performance index (SC_{PI}) are as shown in Equation (6) Eq. (6).

$$SC_{PI} = \frac{SC_{AWC}}{SC_{PWC}} \quad (6)$$

where: SC_{PI}: Screens performance index, SC_{AWC}: Screens actual water cost and SC_{PWC}: Screens planed water cost.

The presented equations for Intake pumps performance index (IP_{PI}), are as shown in Equation (7).

$$IP_{PI} = \frac{IP_{AWC}}{IP_{PWC}} \quad (7)$$

where: IP_{PI}: Intake pumps performance index, IP_{AWC}: Intake pumps actual water cost and IP_{PWC}: Intake pumps planed water cost.

Pretreatment performance index (PRE_{PI}), considering DAF, media filter, cartridge filter and pretreatment dosing system. The presented equations for PRE_{PI}, are as shown in Equation (8).

$$PRE_{PI} = DAF_{PI} + MF_{PI} + CF_{PI} + RDS_{PI} \quad (8)$$

where: PRE_{PI}: Pretreatment performance index, DAF_{PI}: DAF performance index, MF_{PI}: Media filter performance index, CF_{PI}: Cartridge filter performance index and RDS_{PI}: Pretreatment dosing system performance index.

The presented equations for DAF performance index (DAF_{PI}), are as shown in Equation (9).

$$DAF_{PI} = \frac{DAF_{AWC}}{DAF_{PWC}} \quad (9)$$

where: DAF_{PI}: DAF performance index, DAF_{AWC}: DAF actual water cost and DAF_{PWC}: DAF planed water cost.

The presented equations for Media filter performance index (MF_{PI}) are as shown in Equation (10).

$$MF_{PI} = \frac{MF_{AWC}}{MF_{PWC}} \quad (10)$$

where: MF_{PI}: Media filter performance index, MF_{AWC}: Media filter actual water cost and MF_{PWC}: Media filter planed water cost.

The presented equations for Cartridge filter performance index (CF_{PI}), are as shown in Equation (11).

$$CF_{PI} = \frac{CF_{AWC}}{CF_{PWC}} \quad (11)$$

where: CF_{PI}: Cartridge filter performance index, CF_{AWC}: Cartridge filter actual water cost and CF_{PWC}: Cartridge filter planed water cost.

The presented equations for Pretreatment dosing system performance index (RDS_{PI}), are as shown in Equation (12).

$$RDS_{PI} = \frac{RDS_{AWC}}{RDS_{PWC}} \quad (12)$$

where: RDS_{PI}: Pretreatment dosing system performance index, RDS_{AWC}: Pretreatment dosing system actual water cost and RDS_{PWC}: Pretreatment dosing system planed water cost.

RO skid performance index (SK_{PI}), considering DAF, media filter, cartridge filter and pretreatment dosing system. The presented equations for SK_{PI}, are as shown in Equation (13).

$$SK_{PI} = HPP_{PI} + Mem_{PI} + ERD_{PI} \quad (13)$$

where: SK_{PI}: RO skid performance index, HPP_{PI}: High-pressure pump performance index, Mem_{PI}: Membrane performance index and ERD_{PI}: Energy recovery device performance index.

The presented equations for High-pressure pump performance index (HPP_{PI}), are as shown in Equation (14).

$$HPP_{PI} = \frac{HPP_{AWC}}{HPP_{PWC}} \quad (14)$$

where: HPP_{PI}: High-pressure pump performance index, HPP_{AWC}: High-pressure pump actual water cost and HPP_{PWC}: High-pressure pump planed water cost.

The presented equations for Membrane performance index (Mem_{PI}), are as shown in Equation (15).

$$Mem_{PI} = \frac{Mem_{AWC}}{Mem_{PWC}} \quad (15)$$

where: Mem_{PI}: Membrane performance index, Mem_{AWC}: Membrane actual water cost and Mem_{PWC}: Membrane planed water cost.

The presented equations for ERD_{PI}, are as shown in Equation (16).

$$ERD_{PI} = \frac{ERD_{AWC}}{ERD_{PWC}} \quad (16)$$

where: ERD_{PI}: Energy recovery device performance index, ERD_{AWC}: Energy recovery device actual water cost and ERD_{PWC}: Energy recovery device planed water cost.

Post-treatment performance index (PO_{PI}), considering Calcium contactor, post-treatment dosing system and product pumps. The presented equations for PO_{PI}, are as shown in Equation (17).

$$PO_{PI} = CC_{PI} + PDS_{PI} + PP_{PI} \quad (17)$$

where: PO_{PI}: Post-treatment performance index, CC_{PI}: Calcium contactor performance index, PDS_{PI}: Post-treatment dosing system performance index, and PP_{PI}: Product pumps performance index.

The presented equations for Calcium contactor performance index (CC_{PI}), are as shown in Equation (18),

$$CC_{PI} = \frac{CC_{AWC}}{CC_{PWC}} \quad (18)$$

where: CC_{PI}: Calcium contactor performance index, CC_{AWC}: Calcium contactor actual water cost and CC_{PWC}: Calcium contactor planed water cost.

The presented equations for post-treatment dosing system performance index (PDS_{PI}), are as shown in Equation (19).

$$PDS_{PI} = \frac{PDS_{AWC}}{PDS_{PWC}} \quad (19)$$

where: PDS_{PI}: Post-treatment dosing system performance index, PDS_{AWC}: Post-treatment dosing system actual water cost and PDS_{PWC}: Post-treatment dosing system planed water cost.

The presented equations for Product pumps performance index (PP_{PI}), are as shown in Equation (20).

$$PP_{PI} = \frac{PP_{AWC}}{PP_{PWC}} \quad (20)$$

where: PP_{PI}: Product pumps performance index, PP_{AWC}: Product pumps actual water cost and PP_{PWC}: Product pumps planed water cost.

8. Performance Index Calculation Tool

This section describes the creation of a novel analysis tool for helping owners and investors understand the differences between expected and realized water costs, as well as determining the necessary corrective actions. A logic flow chart indicates how we can follow the reason or reasons for the variation depending on checking the difference in the cost of each item, which directs us to the technical reason if applicable. For example, when the tool finds actual power consumption over the planned numbers, it automatically cross-compares actual and planned levels of consumption for all items that were previously highlighted.

A logic flow chart is a step-by-step guideline for following the reasoning for differences by analyzing differences in the cost of individual items. The process leads the analyst to determine any relevant technical reasons for the noted differences. For example, after the tool recognizes that actual power consumption exceeds planned levels, it performs a comparative analysis between actual and planned levels of consumption of all the items already singled out. Such a comparison would allow the operator to target particular components, such as a filter feed pump, which may be using more power than expected, thus enabling the operator to take necessary steps to curb the higher use of power.

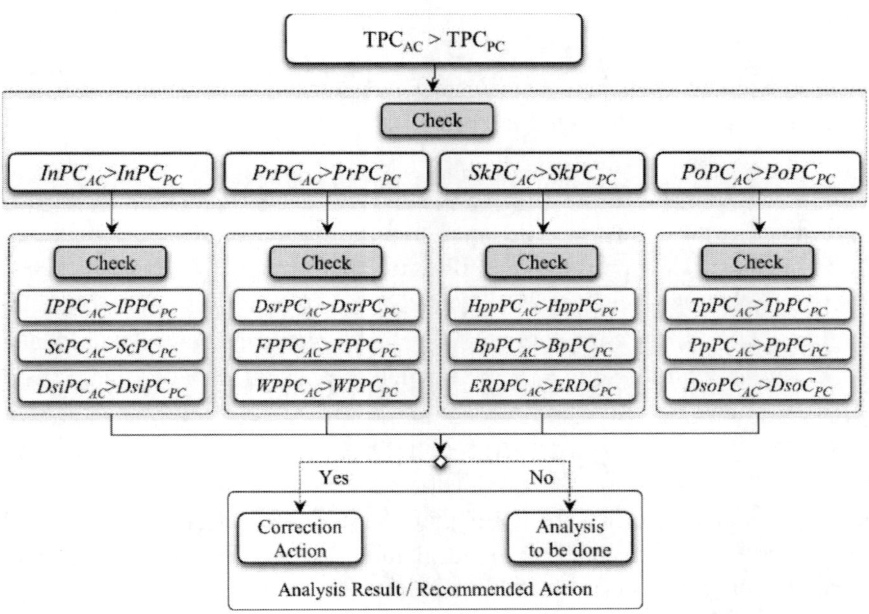

Figure 19. An Example for Total Power Consumption Logic Flow Chart for the Proposed Desalination System.

The Performance Index tool has been developed to reduce the time required by operators and managers to find the causes of increasing operating costs. The proposed logic flowchart displays a methodical way to find out why costs vary by looking at the differences between each cost type. An example of a total power consumption logic flow chart for the proposed desalination plant is shown in Figure 19.

The proposed calculation tool varies depending on the specific case. A simple example of the tool for the total power consumption for a desalination plant consists of four main sections. Power consumption for the intake section (In_{PC}) includes intake pump power consumption (IP_{PC}), screen power consumption (Sc_{PC}), and dosing system power consumption (Dsi_{PC}). Power consumption for the pretreatment section (Pr_{PC}) includes dosing system power consumption (Dsr_{PC}), feed pump power consumption (FP_{PC}), and backwash pump power consumption (WP_{PC}). Power consumption for the RO skid section (Sk_{PC}) includes high-pressure pump power consumption (Hpp_{PC}), booster pump power consumption (Bp_{PC}), and energy recovery device power consumption (ERD_{PC}). Power consumption for the post-treatment section (Po_{PC}) includes transfer pump power consumption (TpPC), product pump power consumption (Pp_{PC}), and dosing system consumption (Dso_{PC}).

9. Performance Index Indicators

A suggested indicator for measuring the Reverse Osmosis Desalination status is presented, which illustrates the operating performance of the desalination plant by RO_{PI}. The indicator classifies the performance index into three different cases, with the first case being $RO_{PI} < 1$ which means that *ATWC* is less than *PTWC*, the second case is $RO_{PI} = 1$, which means that *ATWC* equals *PTWC*, and the third case is $RO_{PI} > 1$, which means that *ATWC* is more than *PTWC*, as shown in Figure 20.

In this scenario, where *ATWC* is initially lower than *PTWC*, it becomes clear that the plant performance indicator is skewed to its advantage, given that the incurred expenses are lower than the budgeted levels. This may, to some degree, be referred to as an ideal indicator by everyone concerned. Technically, two cases arise: either the plant works well with a perfect design or it does not function as per its intended design parameters, and this might even threaten its operational life. Thus, a comprehensive technical appraisal is called for. The equality of *ATWC* and *PTWC* represents an ideal case, which means that everything is running smoothly. However, it is always important to undertake a thorough analysis of all relevant technical and financial data. If *ATWC* exceeds *PTWC*, the difference between them shows potential issues and a detailed technical and financial analysis must be performed. It is also important to set up a margin for the overall water cost for both minimal and maximal values.

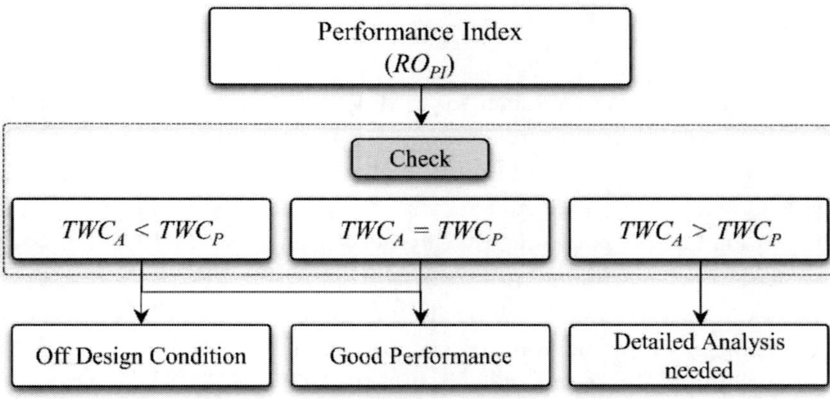

Figure 20. Performance Index Indicators Cases.

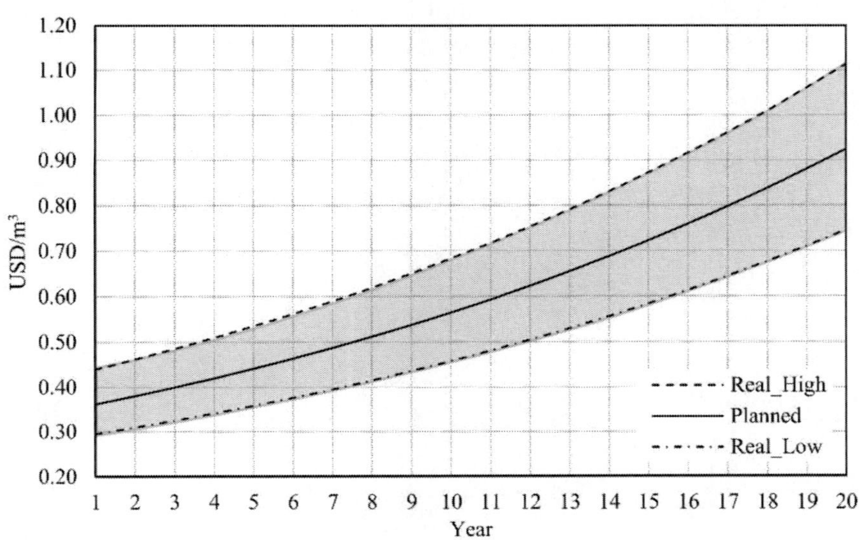

Figure 21. High and Low Total Water Cost for OpEx.

Such a need arises because the operation and maintenance cost calculations for RO systems are influenced by an immense number of factors and parameters. Within the scope of the present chapter, we will only be interested in this margin on the OpEx which is set to ± 10% of the predicted operational costs as Figure 21 shows.

Conclusion

RO_{PI} serves as a critical parameter for facility owners and operators to assess the effectiveness of an integrated reverse osmosis system. The system is comprised of numerous pieces of equipment and components, each exhibiting individual performance characteristics. A significant challenge is the determination of a comprehensive index that accurately reflects the operational capacity of the overall system.

RO_{PI} can be estimated by calculating the ratio of AC to PC to generate one cubic meter of output from the reverse osmosis system. Additionally, a design performance indicator pertaining to the RO system, viz., RO_{DPI}, may be employed to value and evaluate the efficiency of the plant design.

The implementation of the performance index tool is likely to enhance operational effectiveness, thereby allowing the managers and operators to pinpoint the root causes of rising operational expenditures.

Chapter 4

Desalination Systems Environmental Performance Index

Abstract

The environmental impacts resulting from desalination processes are a growing concern, as these technologies involve high energy consumption, High Greenhouse Gases (GHG) emissions, and disturbance to marine ecosystems, primarily through brine disposal and chemicals used during pretreatment. In addition, terrestrial environmental pollution is related to membrane elements, cartridge filters, sports parts, and chemicals. This chapter aims to look into the Environmental Performance Index for Reverse Osmosis Facilities (RO_{EPI}). This will help people in the industry and people who make decisions fully understand the effects on the environment and come up with beneficial ways to lessen those effects. Through the fostering of standards for sustainable operations, this program seeks to harmonize the urgent need for desalinated water with the critical goal of preserving ecosystems and responding to climate change.

Keywords: desalination, reverse osmosis, performance index, desalination environmental performance index

1. Introduction

The supply of freshwater on the planet is basically finite and is being strongly depleted owing to various anthropogenic activities. Consequently, desalination has become a common global method for producing freshwater through non-traditional sources. However, human-engineered technologies are frequently associated with negative environmental impacts (NEI), and this raises concerns about harmful effects. Despite the presence of a reliable source of water, the potential adverse effects of desalination on indigenous freshwater ecosystems remain a vital topic of discussion [83].

The influence of desalination technologies on the environment is highly variable. More specifically, the fast growth of industries like desalination processes leads to more energy use and carbon emissions, which worsens environmental damage and makes global warming worse [84]. Desalination technologies are recognized for their significant energy consumption and the resultant emissions of GHGs. On the other hand, eco-friendly desalination technologies can potentially reduce environmental issues, including climate change, and thus promote sustainable development [74].

For over half a century, RO has been used as a known technique for converting saline water to drinking water and is still the most widely used desalination technology in the world. Recent statistics show that nearly 15,906 operational desalination plants produce about 95 million cubic meters of desalinated water every day for human consumption, with 48% of this production coming from the Middle East and North Africa region [85]. The TDS level of the feedwater has a direct effect on how much energy is used in RO processes. For example, pump pressures between 140 and 400 psi are needed to desalinate brackish water, while pressures as high as 1200 psi are needed to desalinate seawater. This relationship significantly influences both the energy usage and the economic impact of the water produced.

Another big problem with RO systems is that they need chemicals, like chlorine, to get rid of the oxidizing agents in the water that come in before it is treated. For instance, soda fumes and carbon dioxide can be dosed into the water to decrease its acidity, which can then, in turn, impact the performance of the membrane. One of the biggest challenges facing desalination is its environmental impact. The direct and indirect effects on the environment come from different parts of membrane-based desalination, such as the source of the feedwater, the disposal of the brine, the desalination processes, the pretreatment of the feedwater, and the amount of energy needed [86]. One of the major environmental problems is the release of brine, which is made up of liquid wastes from the desalination process, such as chemicals used for pretreatment, corrosive agents, and filter backwash.

As the desalination plants using RO technology are operating, there is a strong need to set up a reference point for the performance efficiency of these facilities. All stakeholders involved in the desalination industry, including manufacturers, solution providers, and operational staff, must adhere to this standardized reference. It is important to be aware that the process of desalination by RO has environmental impacts. For example, the building and running of desalination plants can pollute marine and land ecosystems, cause GHGs emissions to pollute the air, and require a lot of energy.

To help a full study of how well RO systems protect the environment, it is important to create a rule that lets governments, international groups, and people in the industry look at their work as a whole. The goal of this chapter is to introduce a RO_{EPI} that is designed specifically for reverse osmosis desalination facilities. This will help policymakers figure out what steps can be taken to reduce the negative effects of their operations on the environment.

2. General Background

The desalination of water has several detrimental repercussions on the environment, including marine, atmospheric, and terrestrial ecosystems. Moreover, desalination systems have several operational challenges, including corrosion, scale, membrane fouling, and sedimentation. The disposal and treatment of rejected brine from coastal and inland desalination facilities are critical issues for marine ecosystem health and groundwater quality.

3. The Influence of Desalination Plants on the Marine Eco-System

3.1. Feed Water Intake

Desalination techniques of seawater tend to rely on feedwater coming from various sources: open water intakes and subsurface intakes [87]. From among them, open seawater intakes are very commonly used and are likely regarded as the most pragmatic choice. Open intakes, on the other hand, are thought to put aquatic organisms at severe risk, which could harm the ecosystem's balance and cause serious environmental damage. The entrainment, impingement, and entrapment of marine organisms in seawater intake systems pose significant environmental concerns [88-91]. Various aquatic organisms may become impinged on intake screens, entrained into the facility, or entrapped within the intake system. The species encompass algae, fish, jellyfish, shellfish, plankton, and fish eggs [22].

The processes of entrainment and impingement reduce feedwater flow and diminish ecosystem diversity and abundance in terms of species and population numbers. On the other hand, the use of subsurface intakes can perhaps be one of the better mitigation strategies to reduce the harmful impacts of intake systems on marine populations [92].

In addition, intake structures can have an adverse effect on water quality since the construction process may disturb the seabed and thus cause resuspension of pollutants, nutrients, and sediments. The same might happen with debris and dead biomass that should be disposed of after blocking screens and pipelines, which result from entrainment and impingement [87].

3.2. Effluent Discharges

3.2.1. Salinity

The salinity of the reject stream of seawater reverse osmosis (SWRO) desalination facilities was usually about two times higher than that of ambient seawater (about 35,000 ppm). Such a high-salinity and high-temperature reject stream would eventually change the types and numbers of marine life in the area near the discharge site in a big way [93].

High levels of salinity can extend as far as 5 km from the discharge point, thus posing potential threats to a wide array of marine ecosystems, including green seaweed, coral reefs, and seagrass beds [94-96]. The ecological effects of elevated salinity mostly rely on the hydrogeological and environmental features of the specific location. Recent research suggests that hydrodynamic dynamics, rather than hyper salinity, may be crucial in assessing biological consequences around high-pressure diffusers.

Moreover, the density of the reject stream, exceeding that of typical saltwater, may exacerbate its detrimental impact on local species. The heightened density causes the concentrate to settle and disperse along the bottom in shallow coastal regions, hence possibly jeopardizing benthic ecosystems [97].

3.2.2. Desalination Related Chemicals and Byproducts

Antiscalants are polymer-based chemicals utilized in the initial stage of feed water preparation to inhibit scaling. Common antiscalants employed in desalination plants include phosphonates, polycarbonic acids (such as polymeric acid), sulfuric acid, and polyphosphates [66, 96].

Foam control additives: These are chemical agents that mitigate foaming. Antifoaming agents include polyethylene, polypropylene glycol, fatty acid esters, and fatty acids. Although regarded as safe chemicals and found in minimal concentrations within the brine stream, their biodegradability is limited. This may still result in harm to the marine ecosystem.

Chemicals are employed to remove suspended solids. The pre-treatment process in desalination plants is designed to lower suspended solids (turbidity) and other undesirable materials to acceptable levels. Numerous RO plants utilize coagulants and coagulant aids to eliminate suspended solids through coagulation and media filtration. The backwashing water produced by the media filter includes coagulants and suspended solids. Chemical-laden waste is frequently released into marine environments. The presence of such chemicals in the water can result in cloudiness, and may occasionally lead to a change in color. This obstructs sunlight penetration, ultimately harming or potentially leading to the death of specific marine organisms [98].

Cleaning Agents Various cleaning chemicals may be discharged into the environment through waste brine. Acidic solutions (pH 2÷3) are typically employed to remove metal oxides or scales, while alkaline solutions (pH 11÷12) are utilized for membrane cleaning in RO plants. Cleaning membranes involves the removal of biofilm and silt. In addition to these chemicals, the process may also incorporate certain detergents and oxidants. Oxidizing agents such as chlorine and hydrogen peroxide are commonly employed for the cleaning of RO membranes. Non-oxidizing agents such as glutaraldehyde, formaldehyde, and isothiazole are utilized [99].

3.3. Augmented Energy Consumption

Seawater desalination is recognized as an energy-intensive process, requiring substantial energy input for effective operation, as indicated in [100]. Desalination energy consumption can be categorized into thermal energy or electrical energy sourced from the grid. The increased energy consumption linked to these processes is associated with fossil fuel combustion, electricity generation, and the subsequent release of cooling water and airborne pollutants from power production [101]. The GHG and other air pollutant emissions from the generation of steam or electricity by the combustion of fossil fuels are affecting air quality and increasing the rate of global warming phenomena [102]. In addition, the dependence on fuels for energy generation, besides the risks associated with fuel transportation, adds additional indirect environmental impacts. When these direct and indirect effects are considered in an aggregated way, they may lead to cumulative environmental impacts.

3.4. Air Environment Pollution Index (RO_{AEPI})

The air pollution associated with the operation of desalination plants is often significant. When fossil fuels are burned to make steam or electricity, they release GHGs into the air. These include carbon dioxide (CO_2), fine particulate matter (PM), and acid precipitation-forming compounds (including nitrogen oxides [NO_x] and sulfur oxides [SO_x]). These gases, along with many other pollutants, make the air less healthy and speed up global warming. Mostly, there is a direct association of GHG emissions with the use of energy and air pollution.

The magnitude of GHG emissions depends more specifically on the technology of power generation and the type of fuel used; hence, it is contextual. For Multi-Stage Flash (MSF) plants, carbon dioxide emissions range from 9.41 to 25.0 kg CO_2/m^3. For Multi-Effect Distillation (MED) plants, they range from 7.01 to 17.6 kg CO_2/m^3, and for RO plants, they range from 1.75 to 2.79 kg CO_2/m^3. A separate investigation also discovered that an SWRO plant using 4 kWh/m^3 of power will likely release 1.8 kg-CO_2/m^3, 10.7 g-SOx/m^3, 3.9 g-NO_x/m^3, and 2.1 g-$PM10/m^3$ [103].

3.5. Terrestrial Environment Pollution Index (RO_{TEPI})

Desalination plants produce significant solid waste comprising various materials, including membrane elements, cartridge filters, spare parts, and substantial quantities of chemicals, in addition to the reject stream. RO facilities typically produce a greater volume of solid waste than thermal desalination facilities. The environmental impacts of solid waste are significantly influenced by its chemical composition [104, 105].

4. Desalination's Role in Aggravating Climate Change

Desalination is a process that requires significant energy consumption.; most of its energy supply comes from fossil fuels, which are classified as non-renewable energy sources. Consequently, the impacts of climate change pose significant challenges to desalination facilities. Two key issues related to energy consumption in desalination plants should be addressed. First, there is a need to lower total energy use. Second, it is important to use renewable energy sources to mitigate GHG emissions. At the moment, desalination processes pollute the air by releasing carbon oxides (CO_x), nitrogen oxides (NO_x), sulfur oxides (SO_x), and particulate matter. This is because they use a lot of energy and rely on fossil fuels that can't be replaced. Scientists discovered that SWRO desalination processes powered by natural gas released 2.79 kg CO_2/m^3 of produced water for the steam cycle, 2.13 kg CO_2/m^3 of produced water for the internal combustion engine, and 1.75 kg CO_2/m^3 of produced water for the combined cycle.

Moreover, their study calculated the global warming potential of SWRO desalination plants and found an emission value of 1.77 kg CO_2/m^3 of produced water. Acidification potential, in terms of sulfur dioxide (SO2), was also estimated at g-SO2/m^3 of the produced water [106]. A SWRO desalination plant gives 2.3 to 2.5 kg CO_2/m^3 as a carbon footprint, equivalent to a production capacity of 0.63 million cubic meters per day (MCM/day) [100, 107].

5. Objectives

Establishing RO_{EPI} is significant for assessing the negative environmental impacts associated with these operations and mitigating the impacts of climate change. This performance index serves as a quantitative measure that assesses how much the operations of the desalination facility deviate from its pre-established environmental targets. Throughout history, with the widespread use of RO technology, numerous research studies have been conducted to come up with ways to evaluate the performance of such systems. Most of these methods have focused on how well the membrane can make drinking water, pointing out that factors like membrane flux and salt rejection are very important for a RO system to work.

After that, the study grew to include how much energy and chemicals are needed to make one cubic meter of freshwater. These are the most important factors in figuring out how well the RO process works overall [108].

The study looks at how to make an Environmental Performance Index, especially for RO desalination plants (RO_{EPI}), so that the environmental impact of the whole desalination process can be measured. It is important to note that the interplay between technical efficiency and operating costs has a significant impact on the total ecological footprint. A loss in efficiency tends to rise with increased costs and escalated negative environmental impacts; however, attaining optimal efficiency can lead to lower expenses and reduced ecological impacts. As a result, this study stresses how important it is to find the difference between how much was planned to be spent on environmental protection and how much was actually spent to make one cubic meter of water in the whole RO system.

6. Methodology

The Environmental Performance Index (EPI) is a metric that translates the NEI results into production cost objectives for the operational lifespan of a facility. The RO_{EPI} is determined by dividing the Actual Cost of Environmental Evaluation Parameters (EEP_{AC}) by the Planned Cost of Environmental Evaluation Parameters (EEP_{PC}) required for the production of one cubic meter of water in the whole RO system.

The expenses associated with the environmental evaluation criteria of RO desalination facility designed for a particular configuration and location are influenced by a multifaceted set of variables. They include the feed water quality (salinity, turbidity, pH, chemical oxygen demand (COD), and biological oxygen demand (BOD)), feed water temperature, membrane fouling, quality of chemicals, maintenance downtime schedule, and degree of competence of operating personnel.

Since these factors are indeterminate, minimum and maximum figures must be set for the environmental assessment parameters. The budgeted cost is used as the basis to approximate the EPI, as indicated in Figure 22.

Two options may be evaluated for the calculation of the Planned Cost (PC):

- The design parameters were followed.
- The system is functioning in accordance with the startup parameters.

The planned cost (PC) of RO systems is inherently linked to design parameters that account for variations in feed water temperature and membrane age. The variability in planned costs produces a spectrum of potential values. This presents decision-makers with the option to employ minimum, maximum, or average values, or to create a time-dependent function of the planned cost, PC(t).

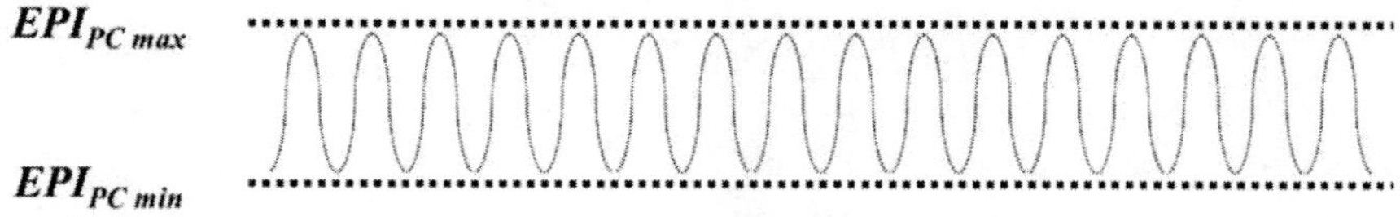

Figure 22. Upper and Lower Environmental Assessment Criteria for Desalination System.

In the context of startup operating conditions, the planned cost (PC) is determined based on actual data regarding feed water temperature, membrane aging, feed water quality, and other relevant parameters at a specific time.

Given the unpredictability of parameter variations under actual operating conditions and their direct impact on costs, a hybrid strategy is proposed. The essential components of the hybrid strategy include:

- Develop a cost graph illustrating production costs as a function of time throughout the plant's lifecycle concerning design parameters, represented as PCD(t).
- Develop a secondary cost graph illustrating production costs throughout the plant's operational lifespan, integrating startup parameters to depict trends relative to design parameters. The graph, designated PCS(t), will serve as a reference point.

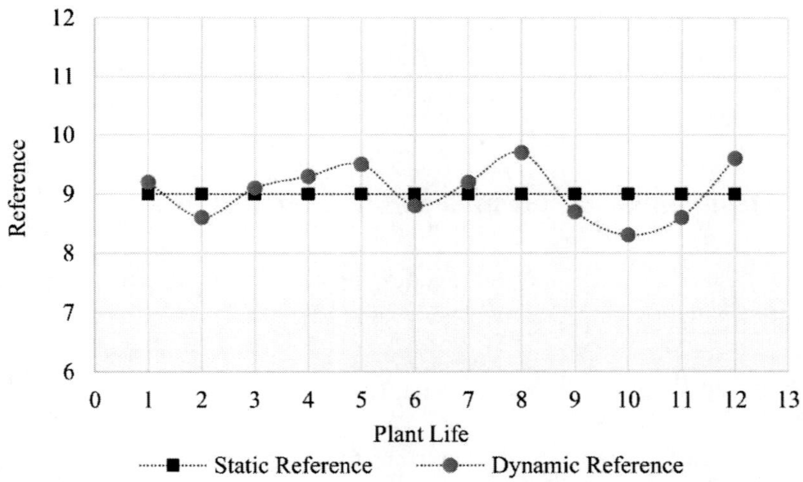

Figure 23. Environmental Performance Index Dynamic Reference.

The operation parameters of the RO system vary over time; thus, Figure 23 illustrates that a dynamic benchmark is the most appropriate method for calculating the Environmental Performance Indicator (EPI) throughout the plant's life cycle. An Environmental Performance Index, termed RO_{EPI}, has the potential to facilitate the measurement and calculation of plant design efficiency based on the design of the RO system.

The Environmental Performance of the overall system relies on the appropriate design and functioning of each component. The RO plant must sustainably generate the necessary quantity of drinkable water of acceptable quality at minimal cost.

7. Environmental Performance Index

The necessity for the assessment of the environmental footprint of RO desalination plants is imperative to reduce their contribution to climate change. The (RO_{EPI}) is used to measure the consequences of both the design and operational parameters of each plant on global warming and climate change. It is necessary to develop a structure that will allow governments, international organizations, stakeholders, and operators to measure the environmental performance of such plants in the context of an integrated system, for example, the RO system, which is made up of a huge number of equipment and components. The objective of this chapter is to create an overall (RO_{EPI}) as a tool to assess their plants and implement corrective actions to minimize environmental effects in a real-case scenario. The suggested RO_{EPI} system is modeled based on three main components, as shown in Figure 24.

8. Environmental Performance Index Calculation

The proposed EPI consists of three main components, as shown in Figure 24 and represented by both Equations (21) and (22).

$$RO_{EPI} = RO_{MEPI} + RO_{TEPI} + RO_{AEPI} \tag{21}$$

where: RO_{EPI}: Reverse osmosis environmental performance index, RO_{MEPI}: Reverse osmosis marine environmental performance index, RO_{TEPI}: Reverse osmosis terrestrial environmental performance index and RO_{AEPI}: Reverse osmosis air environmental performance index.

$$RO_{EPI} = \frac{RO_{EPIAC}}{RO_{EPIPC}} = \frac{RO_{MEPIAC}}{RO_{MEPIPC}} + \frac{RO_{TEPIAC}}{RO_{TEPIPC}} + \frac{RO_{AEPIAC}}{RO_{AEPIPC}} \tag{22}$$

where: RO_{EPI} refers to the Reverse Osmosis Environmental Performance Index. RO_{EPIAC} denotes the actual cost of estimation parameters for the Reverse Osmosis Environmental Performance Index. RO_{EPIPC} indicates the planned cost of these estimation parameters. RO_{MEPIAC} represents the actual cost of estimation parameters for the Reverse Osmosis Marine Environmental Performance Index, while RO_{MEPIPC} signifies the planned cost. RO_{TEPIAC} refers to the actual cost of estimation parameters for the Reverse Osmosis Terrestrial Environmental Performance Index, and RO_{TEPIPC} indicates the planned cost. Finally, RO_{AEPIAC} denotes the actual cost of estimation parameters for the Reverse Osmosis Air Environmental Performance Index, whereas RO_{AEPIPC} represents the planned cost.

The Marine Environment Pollution Index (RO_{MEPI}) takes into account parameters like elevated salinity, elevated temperature, elevated pH, and chemical residues.

The governing equations for the (RO_{MEPI}) are presented in Equation (23).

$$RO_{MEPI} = RO_{SaEPI} + RO_{TeEPI} + RO_{pHEPI} + RO_{CrEPI} \tag{23}$$

where: RO_{MEPI} refers to the reverse osmosis marine environment pollution index, RO_{SaEPI} denotes the reverse osmosis salinity environment pollution index, RO_{TeEPI} signifies the reverse osmosis temperature environment pollution index, RO_{pHEPI} indicates the reverse osmosis pH environment pollution index, and RO_{CrEPI} represents the reverse osmosis chemical residuals environment pollution index.

The governing equations for Reverse Osmosis Salinity Environment Pollution Index (RO_{SaEPI}) are as shown in Equation (24).

$$RO_{SaEPI} = \frac{Sa_{AC}}{Sa_{PC}} \tag{24}$$

where: RO_{SaEPI} refers to the Reverse Osmosis Salinity Environment Pollution Index, Sa_{AC} denotes the actual cost of salinity environment evaluation parameters, and Sa_{PC} indicates the planned cost of salinity environment evaluation parameters.

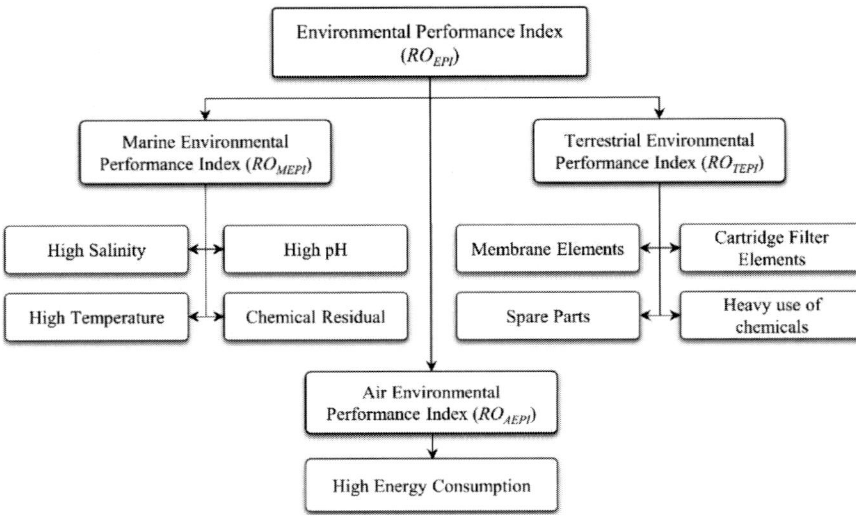

Figure 24. Key Components of the Reverse Osmosis Environmental Performance Index.

The governing equations for Reverse Osmosis Temperature Environment Pollution Index (RO_{TeEPI}) are as shown in Equation (25).

$$RO_{TeEPI} = \frac{Te_{AC}}{Te_{PC}} \tag{25}$$

where: RO_{TeEPI} refers to the reverse osmosis temperature environment pollution index, while Te_{AC} denotes the actual cost of temperature environment evaluation parameters, and Te_{PC} signifies the planned cost of these parameters.

The governing equations for Reverse Osmosis pH Environment Pollution Index (RO_{pHEPI}) are as shown in Equation (26).

$$RO_{pHEPI} = \frac{pH_{AC}}{pH_{PC}} \tag{26}$$

where: RO_{pHEPI} refers to the reverse osmosis pH environment pollution index, pH_{AC} denotes the actual cost of pH environment evaluation parameters, and pH_{PC} indicates the planned cost of pH environment evaluation parameters.

The governing equations for Reverse Osmosis Chemical Residuals Environment Pollution Index (RO_{CrEPI}) are as shown in Equation (27).

$$RO_{CrEPI} = \frac{Cr_{AC}}{Cr_{PC}} \tag{27}$$

where: RO_{CrEPI}refers to the reverse osmosis chemical residuals environment pollution index, while Cr_{AC}denotes the actual cost of chemical residuals environment evaluation parameters, and Cr_{PC}signifies the planned cost of these parameters.

The governing equations for Reverse Osmosis Marine Environment Pollution Index (RO_{MEPI}) are as shown in Equation (28).

$$RO_{MEPI} = \frac{Sa_{AC}}{Sa_{PC}} + \frac{Te_{AC}}{Te_{PC}} + \frac{pH_{AC}}{pH_{PC}} + \frac{Cr_{AC}}{Cr_{PC}} \tag{28}$$

where: RO_{MEPI} refers to the reverse osmosis marine environment pollution index. Sa_{AC} denotes the actual cost of salinity environment evaluation parameters, while Sa_{PC} indicates the planned cost for the same parameters. Te_{AC} represents the actual cost of temperature environment evaluation parameters, and Te_{PC} signifies the planned cost. pH_{AC} is the actual cost associated with pH evaluation environment parameters, whereas pH_{PC} refers to the planned cost. Lastly, Cr_{AC} pertains to the actual cost of chemical residuals environment evaluation parameters, and Cr_{PC} denotes the planned cost for these parameters.

Terrestrial Environment Pollution Index (RO_{TEPI}), regarding membrane elements, cartridge filter elements, spare parts, and extensive chemical usage. The governing equations for the Terrestrial Environment Pollution Index (RO_{TEPI}) are presented in Equation (29).

$$RO_{TEPI} = RO_{MeEPI} + RO_{CaEPI} + RO_{SpEPI} + RO_{ChEPI} \tag{29}$$

where: RO_{TEPI} refers to the reverse osmosis terrestrial environment pollution index, RO_{MeEPI} denotes the reverse osmosis membrane elements environment pollution index, RO_{CaEPI} indicates the reverse osmosis cartridge filter element's environment pollution index, RO_{SpEPI} represents the reverse osmosis spare parts environment pollution index, and RO_{ChEPI} signifies the reverse osmosis chemical environment pollution index.

The governing equations for Reverse Osmosis Membrane Elements Environment Pollution Index (RO_{MeEPI}) are as shown in Equation (30).

$$RO_{MeEPI} = \frac{Me_{AC}}{Me_{PC}} \tag{30}$$

where: RO_{MeEPI} index for environmental pollution associated with reverse osmosis membrane elements. Me_{AC} Actual cost of membrane environment evaluation parameters and Me_{PC} Planned cost of membrane environment evaluation parameters.

The governing equations for Reverse Osmosis Cartridge Filter Elements Environment Pollution Index (RO_{CaEPI}) are as shown in Equation (31).

$$RO_{CaEPI} = \frac{Ca_{AC}}{Ca_{PC}} \tag{31}$$

where: RO_{CaEPI} Environmental pollution index of reverse osmosis cartridge filter elements. Ca_{AC} Actual cost of cartridge filter elements environmental evaluation parameters and Ca_{PC} Planned cost of cartridge filter elements environmental evaluation parameters.

The governing equations for Reverse Osmosis Spare Parts Environment Pollution Index (RO_{SpEPI}) are as shown in Equation (32).

$$RO_{SpEPI} = \frac{Sp_{AC}}{Sp_{PC}} \tag{32}$$

where: RO_{SpEPI} refers to the reverse osmosis spare parts environment pollution index, while Sp_{AC} denotes the actual cost of spare parts environment evaluation parameters, and Sp_{PC} indicates the planned cost of these parameters.

The governing equations for Reverse Osmosis Chemical Environment Pollution Index (RO_{ChEPI}) are as shown in Equation (33).

$$RO_{ChEPI} = \frac{Ch_{AC}}{Ch_{PC}} \tag{33}$$

where: RO_{ChEPI} refers to the Reverse Osmosis Chemical Environment Pollution Index, while Ch_{AC} denotes the actual cost of chemical environment evaluation parameters, and Ch_{PC} indicates the planned cost of these parameters.

The governing equations for Reverse Osmosis Terrestrial Environment Pollution Index (RO_{TEPI}) are as shown in Equation (34).

$$RO_{TEPI} = \frac{Me_{AC}}{Me_{PC}} + \frac{Ca_{AC}}{Ca_{PC}} + \frac{Sp_{AC}}{Sp_{PC}} + \frac{Ch_{AC}}{Ch_{PC}} \quad (34)$$

where: RO_{TEPI} refers to the reverse osmosis terrestrial environment pollution index. Me_{AC} denotes the actual cost of membrane environment evaluation parameters, while Me_{PC} indicates the planned cost for these parameters. Ca_{AC} represents the actual cost of cartridge filter elements environment evaluation parameters, and Ca_{PC} signifies the planned cost. Sp_{AC} is the actual cost for spare parts environment evaluation parameters, with Sp_{PC} reflecting the planned cost. Lastly, Ch_{AC} indicates the actual cost of chemical environment evaluation parameters, and Ch_{PC} denotes the planned cost.

Air Environment Pollution Index (RO_{AEPI}), relation to elevated energy consumption. The governing equations for Air Environment Pollution Index (RO_{AEPI}) are presented in Equation (35).

$$RO_{AEPI} = RO_{EnEPI} \quad (35)$$

where: RO_{MEPI} refers to air environment pollution index, while RO_{EnEPI} denotes Energy environment pollution index.

The governing equations for Reverse Osmosis Energy Environment Pollution Index (RO_{EnEPI}) are as shown in Equation (36).

$$RO_{EnEPI} = \frac{Energy_{AC}}{Energy_{PC}} \quad (36)$$

where: RO_{EnEPI} refers to the Energy Environment Pollution Index, while $Energy_{AC}$ denotes the actual cost of energy environment evaluation parameters, and $Energy_{PC}$ represents the planned cost of these parameters.

9. Environmental Performance Index Calculation Tool

An envisioned analytical instrument is crucial for identifying disparities between anticipated and actual environmental consequences, thus assisting owners and investors in selecting requisite remedial measures. Analyzing the cost differentials of each component allows for the identification of any underlying technical explanations for these discrepancies. If the tool indicates that real power consumption surpasses anticipated estimates, it will methodically analyze the actual and planned energy use for all components recorded inside the tool. Consequently, the operator will be informed that the power consumption of the filter feed pump is much greater than originally projected. This compels the operator to execute remedial actions to rectify the increased power usage. A segment of the relevant logic flow chart is displayed in Figure 25.

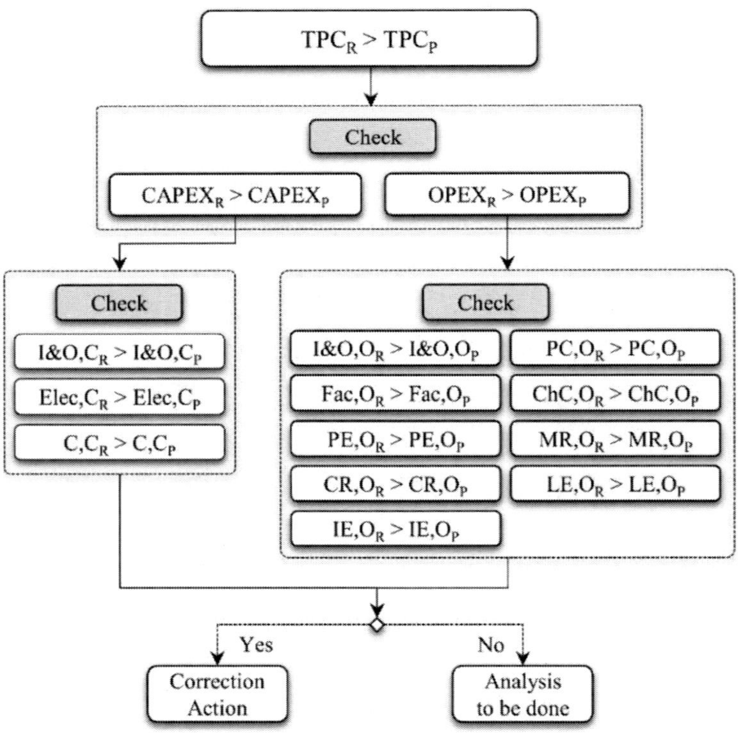

Figure 25. Flowchart for the Calculation of the Environmental Performance Index.

10. Environmental Performance Index Indicators

An indicator for the Environmental Status of Reverse Osmosis Desalination is developed, which reflects the plant's environmental condition based on the RO_{EPI}.

This indicator demonstrates that the RO_{EPI} has three cases; the initial instance is (RO_{EPIAC}) < (RO_{EPIPC}), the second case is (RO_{EPIAC}) = (RO_{EPIPC}) and the third case is (RO_{EPIAC}) > (RO_{EPIPC}), as illustrated in Figure 26.

In the case of the indication (RO_{EPIAC}) being less than (RO_{EPIPC}), The initial assessment indicates that the environmental plant performance index is favorable, since the costs are lower than anticipated. early evaluations can indicate a satisfactory environmental performance of the facility, as operational costs are seemingly less than projections.

Technically, this situation can imply one of two things: Either the facility is showing optimum environmental efficiency due to a well-planned design, or it may be operating outside the intended parameters, hence undermining its durability. A comprehensive technical assessment is thus warranted.

Equality of the indices (RO_{EPIAC}) and (RO_{EPIPC}) is the ideal case, which means that the plant is running at its optimum. Nevertheless, it is necessary to carry out a detailed analysis of all the pertinent technical and financial information to validate this conclusion.

Conversely, a situation where (RO_{EPIAC}) exceeds (RO_{EPIPC}) signals potential concerns regarding the plant's performance, necessitating an in-depth technical and financial examination. A critical examination of all the environmental factors will facilitate the most effective interpretation of the data and determine particular factors that can adversely affect the plant's performance. Figure 27 to Figure 30 displays the systematic review process of this analysis. The (RO_{EPI}) is a key indicator for assessing the environmental impacts of RO plants and their effects on climate change.

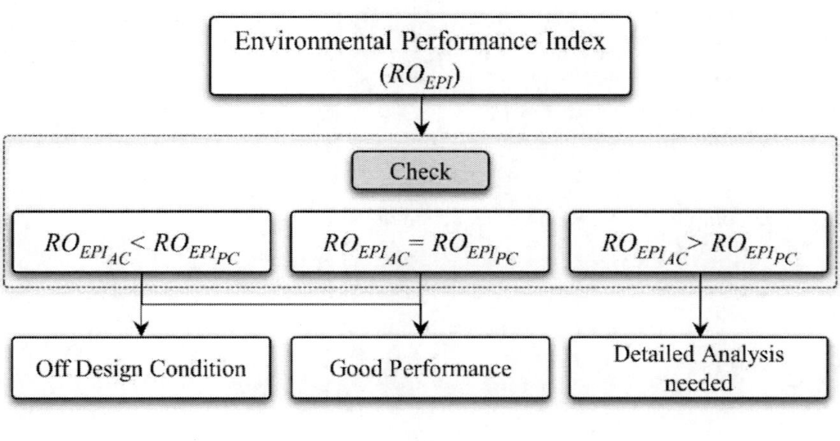

Figure 26. Environmental Performance Index Indicators Cases.

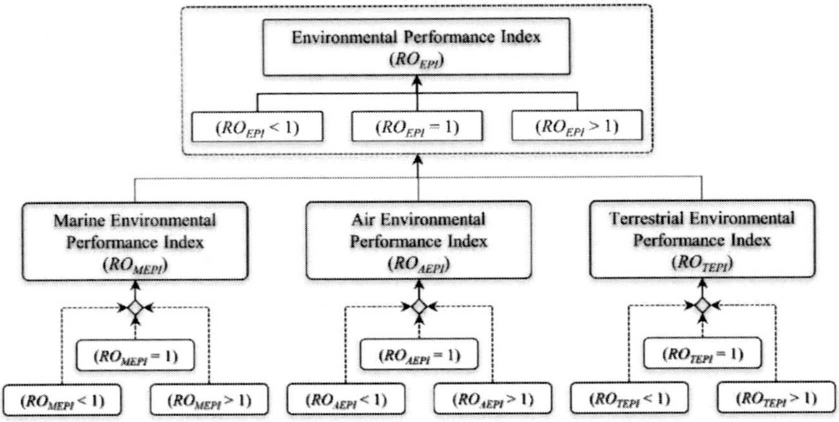

Figure 27. Comprehensive Review Sequence of Environmental Performance Index Main Items.

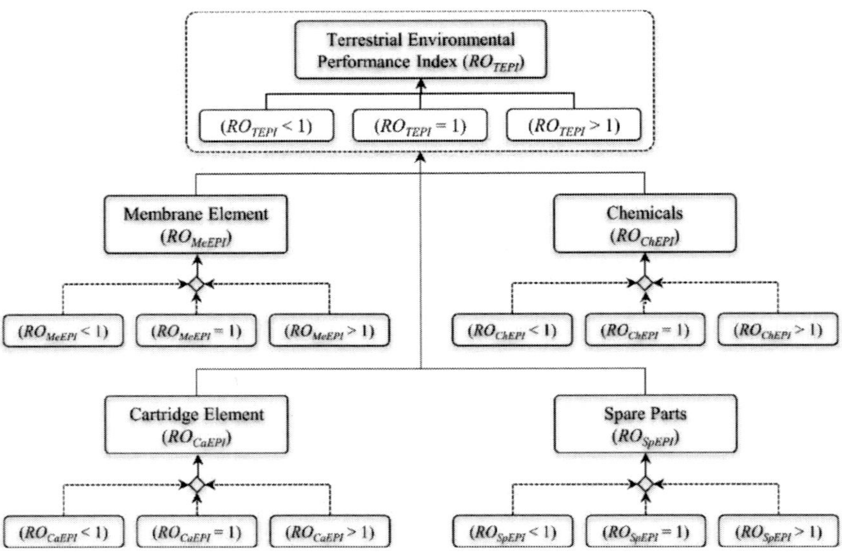

Figure 28. Comprehensive Review Sequence of Terrestrial Environmental Performance Index Items.

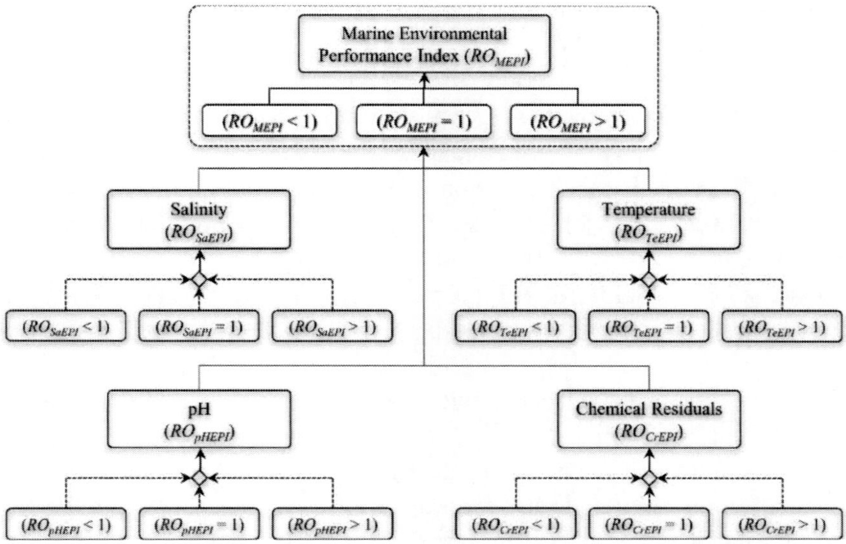

Figure 29. Comprehensive Review Sequence of Marine Environmental Performance Index Items.

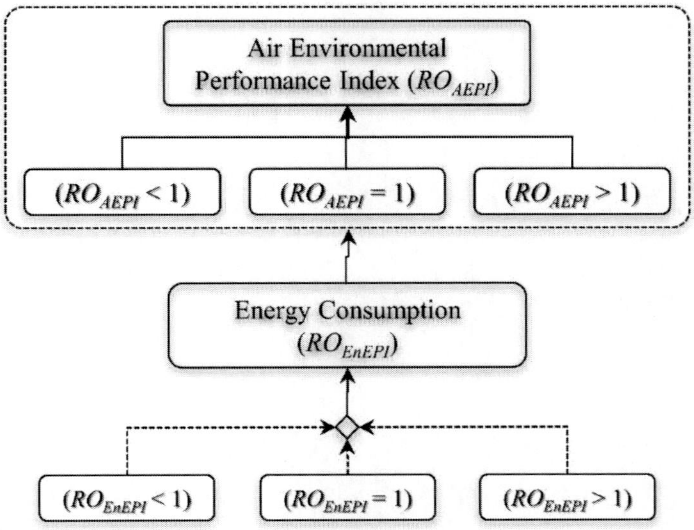

Figure 30. Comprehensive Review Sequence of Air Environmental Performance Index Items.

Conclusion

Although desalination, especially using reverse osmosis technology, is a viable solution to the increasing need for freshwater, its effects on the environment must be recognized and alleviated. In addition to raising energy use and GHGs emissions, desalination also causes serious harm to marine and freshwater ecosystems in several ways, including brine discharge and aquaculture intake.

Creating a standardized tool like the (RO_{EPI}) for desalination plants can be a significant step toward encouraging sustainable practices in the industry. By involving all stakeholders, from manufacturers to policymakers to plant operational personnel, we can strive towards reducing the environmental footprint of desalination and making clean water available for generations to come. It's essential that we accord the highest priority to both desalination technology innovation and the health of our natural systems so that we can have a balanced approach to water resource management.

Chapter 5

Artificial Intelligence and Simulation Integration for Sustainable Desalination

Abstract

This chapter introduces a concise examination of Artificial Intelligence (AI) as a significant technological innovation, along with simulation and machine learning discussions. Different definitions of AI and simulation models are looked at to get a full picture of the term. These definitions are based on how autonomous they are and how well they can improve industry performance and process understanding. The evolution of these two concepts is traced from their beginnings in the mid-20th century through to their current advancements motivated by big data and machine learning technologies. The chapter sorts AI, machine learning, and simulations into groups based on their abilities and stresses how AI systems can do many things. Lastly, the discussion introduces the capability of AI to improve operational effectiveness, minimize human error, and address intricate problems in most industries, especially in terms of creating innovative solutions for desalination.

Keywords: artificial intelligence, history of artificial intelligence, simulation integration, desalination and simulation. sustainable desalination, machine learning

1. Introduction

The desalination process is known for having a lot of different complicated steps. These steps include choosing the right site, figuring out the best size, and setting up the operational parameters. Each of these steps has its own set of complicated nonlinear problems and different levels of freedom, whether the plant is being designed, built, or maintained. Parameter selection is typically done depending to a large extent on the experience and intuition of researchers and engineers. Furthermore, the design of large desalination plants entails a vast number of design parameters and requirements, leading to a complex integrated design problem. The presence of uncertain factors, i.e., time-varying loads and energy inputs, makes the situation more challenging, as there is no single, always-optimal solution that guarantees system optimization. The trade-off between the two objectives of cost and efficiency is troublesome based on the expertise and judgment of experts alone. The above difficulties impede the progress of mega seawater desalination projects. Because of this, institutions and researchers are becoming more and more interested in making technologies that can achieve optimal regulation on several dimensions while also reducing the need for human labor and speeding up complex calculations [109-112].

With the evolution of the information era in modern times, the advancement of network technologies has encouraged various entities to start programs aimed at ensuring the best industrial advancement. Recent studies often incorporate machine learning and AI into the platform of energy system planning. Computing technologies are renowned for their explanation, flexibility, and semiotic reasoning capabilities in engineering applications. The use of AI technology does not take away the operator's control or involvement in manual tasks; instead, it enhances system performance by providing fast and accurate responses from computing systems, leading to better quality and more efficient production. AI possesses the ability to augment overall process efficacy and minimize human error; AI is superior in intelligent decision-making and evaluation.

AI is also a cutting-edge engineering technology that can be used to help desalinate seawater. This capability frees up human operators to focus on the complicated calculations, design improvements, and control needs of these processes [113]. The use of artificial in AI in the desalination of seawater began in the 1990s. Since then, several studies have considered the use of computational intelligence (CI) systems in the field. CI is mostly used in three areas in desalination: prediction (like load forecasting), control (like handling alarms and finding faults), and global evaluation (like operational optimization, security analysis, and economic analysis). CI, in essence, is an advanced computer program designed to assist individuals in resolving various problems based on independent thinking [31]. Some of the researchers have focused their efforts on improving the mathematical underpinnings of intelligent algorithms utilizing advanced computing techniques [114].

Despite these advancements, a thorough framework and architecture for the integration of AI in desalination design has yet to be established. The effectiveness of various methods for prediction and optimization might lead to discrepancies in optimum operations. We need to create a clear plan to include AI in both local and overall desalination processes, as well as ways to use smart algorithms [115]. To achieve an effective description and analysis of the desalination process, researchers have been investigating the use of AI techniques and data-driven methodologies, as seen in numerous researches.

Conventional model-based techniques tend to suffer from the complexity of desalination and are thus unsatisfactory. AI-based models, on the other hand, demonstrate proficiency in translating large volumes of data into valuable information and precise behavioral models. This capability highlights the key advantages of applying AI techniques, which are essential for modeling, optimizing, and controlling a broad array of applications [116, 117].

2. Artificial Intelligence Overview

AI can be termed the most impactful technological development of the past decade, and it will continue to shape future advancements. This chapter focuses on AI functions and their applicability, more so in the context of desalination processes. There have been multiple definitions of AI from different points of view, showing diversity in aspects of the field. Some of these definitions will be examined. Hence, in this chapter, we introduce a definition that covers the AI concept. AI just refers to systems that can carry out intelligent behavioral reactions by recognizing their environment. Such systems have some degree of autonomy in their actions since they try to accomplish some form of goal.

AI is probably the defining technology of the last decade and most probably the next. An accessible overview of how AI works will be introduced and how it matters, especially in desalination. Many definitions of AI have been proposed based on experts' points of view; some of the definitions will be presented, followed by our preferred definition. AI refers to systems that can introduce intelligent behavior reactions by analyzing their environment and acting with some degree of autonomy to achieve specific goals. AI is likely the paramount technology of the last decade and perhaps the next one. An understandable introduction to the functioning of AI will be presented, highlighting its significance, particularly in desalination.

A multitude of definitions of AI have been given, reflecting the perspectives of many specialists; several of these definitions will be provided, followed by our recommended definition. AI denotes systems capable of exhibiting intelligent behavior by assessing their surroundings and operating with a certain level of autonomy to accomplish designated objectives [118]. You can think of AI as a computer program that tries to think like humans. Its main goal is to achieve specific objectives within a defined environment with a degree of autonomy [118]. In addition, it can be described as machines mimicking human mental processes, especially in computing.

In the same way, AI can be thought of as a system created by humans that can perform intelligent tasks autonomously [119]. The capacity of machines to emulate intelligence is what is well known about AI a set of algorithms that can have the capability to produce outputs without explicit directions [120]. For this purpose, we give the following definitions: AI is the emulation of natural intelligence in machines to acquire knowledge and mimic human action. Despite the different connotations assigned to AI, its core purpose is to increase efficiency and restrict human mistakes in different aspects of life.

This is achieved through a simulation of brain functions, thus leading to technologies that aid reasoning and decision-making [121]. In some cases, the use of AI technologies can not only solve multifaceted and difficult problems in different fields, like those related to desalination, but they can also worsen them. They provide the means for designing new and effective solutions [122].

AI is an advanced computer technology that is designed for performing tasks that would otherwise require human intellectual abilities. However, AI systems can learn from their mistakes and improve their performance and efficiency over time, unlike automated equipment and machines that follow set rules and always give the same answer [123].

2.1. Artificial Intelligence History

The history of AI began in the mid-20th century, when the concept of "thinking machines" began to gain momentum. Some of the milestones in the evolution and growth of AI are mentioned below:

- Early Concepts (1940s-1950s): Computer scientists and mathematicians began investigating AI's underlying concepts, most prominently Alan Turing, who proposed the Turing Test as a metric to evaluate the possibility of a machine replicating intelligent behavior.
- Dartmouth Conference (1956): The Dartmouth College conference, which was organized by key players like John McCarthy, Marvin Minsky, Nathaniel Rochester, and Claude Shannon, is where the term "Artificial Intelligence" was formally used. This conference is

generally regarded as the beginning of AI as a serious field of academic research.

- Early Development (1950s-1970s): Initial research focused on developing algorithms and programming languages, such as LISP, to facilitate AI experimentation. Early AI systems included simple problem-solving programs and rule-based systems [124].
- AI Winter (1970s-1980s): During this period, enthusiasm for AI waned due to unmet expectations and limited computational power. Funding and interest declined, leading to what is known as the "AI Winter" [125].
- Renewed Interest (1980s-1990s): Advances in machine learning, neural networks, and expert systems reignited interest in AI. The development of more powerful computers and the availability of large datasets contributed to this resurgence.
- The years between the 2000s till now are a single leap in the history of AI, fueled mostly by the advent of big data, the creation of deep learning algorithms, and the application of complex algorithms. Some of the advancements that have been made over the period include the creation of natural language processing systems like GPT-3, the creation of computer vision algorithms, and the spread of autonomous systems [126].

2.2. Artificial Intelligence Types

AI can be organized systematically based on its various capabilities and functionalities. A comprehensive knowledge of such categorizations is extremely crucial concerning understanding both the present state and future development of AI technologies, illustrated in Figure 31.

2.2.1. AI Types Based on Capabilities

AI has three types based on the criteria of capability: artificial narrow intelligence (Narrow AI), general artificial intelligence (General AI), and artificial superintelligence (Super AI).

Artificial Narrow Intelligence, is linked to AI systems that are specifically designed to perform predefined tasks or instructions. These systems are characterized by their competence in a particular cognitive domain and lack of ability to develop skills independently beyond their programming. All AI systems implemented in real world are narrow AI systems. This type of intelligence typically employs methods such as machine learning, natural language processing, and neural network-based algorithms to carry out given tasks. A few of the well-known examples of narrow AI include autonomous vehicles and AI-powered virtual assistants [127].

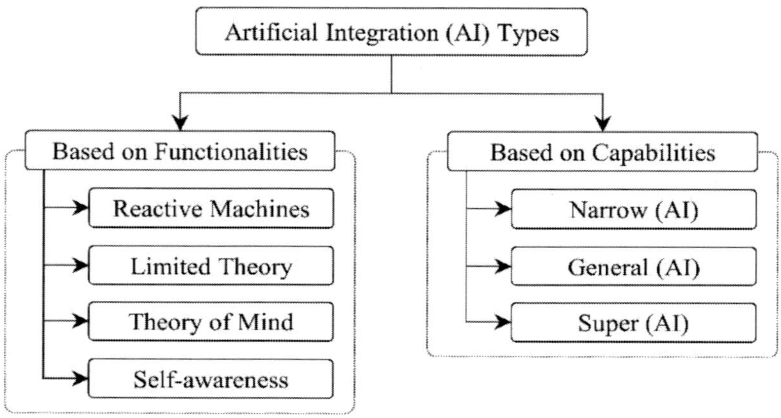

Figure 31. Artificial Intelligence Types Based on Capabilities and Functionalities.

Artificial General Intelligence (AGI): developed to acquire knowledge, reason, and operate at levels comparable to humans. AGI, sometimes known as general AI or strong AI, denotes a hypothetical type of AI capable of learning, reasoning, and executing a diverse array of activities at a human level. The primary objective of AGI is to develop robots that possess flexible, human-like intelligence, serving as highly adaptable aides in daily life. The groundwork for artificial general intelligence may be laid, despite being in its nascent phase. The groundwork for artificial general intelligence may be established using technologies such as supercomputers, quantum hardware, and generative AI applications [128].

Artificial Super Intelligence (ASI): capable of exceeding human knowledge and skills. ASI, also referred to as super AI, is an idea similar to science fiction. It is postulated that upon achieving universal intelligence, AI would rapidly acquire knowledge and skills, surpassing human capacities. ASI would serve as the foundational technology for fully self-aware AI and another autonomous robotics. This idea also underpins the prevalent media narrative of "AI takeovers." At this juncture, it is only conjecture. An illustration of super AI is a robot [129].

2.2.2. AI Types Based on Functionalities

Based on functionality, there are four types of AI: Reactive Machines, Limited Memory, Theory of Mind, and Self-awareness.

- Reactive Machine Artificial Intelligence: Reactive machines represent the fundamental building blocks of AI, distinguished by their swift reaction times. Reactive machines are those that are designed to respond to a certain request and task without possessing the ability to store memory, learn from previous interactions, or enhance performance over time depending on previous interactions. In addition, reactive machines have a small number of input combinations, and therefore their activities are limited. Reactive machines perform well in executing necessary autonomous functions in real-world situations, such as blocking spam from email accounts and recommending products based on buying history. However, reactive AI's inherent limitations prevent it from leveraging acquired knowledge or performing more complex tasks [130].
- Limited Memory Artificial Intelligence: Limited memory AI is used to describe systems with the capacity to retain and use past data to improve forecasting and decision-making. This capacity enables these systems to develop a limited, short-term knowledge base used in performing a set of tasks. Deep learning, an approach that mimics the role of neurons in the human brain, is at the center of limited memory AI. This design paradigm allows knowledge to be amassed from numerous experiences, thus leading the machine to "learn" and enhance its work processes over time. Limited memory AI accounts for most applications within the AI industry today, and potential uses in the future range from simple uses such as chatbots to advanced technologies such as self-driving cars [131].
- Theory of Mind Artificial Intelligence: The theory of mind in AI refers to the ability to recognize and interpret the emotions of others. The term is borrowed from psychology, describing humans' ability to read the emotions of others and predict future actions based on that information. Although it has not been achieved yet, theory of mind would be a substantial milestone in AI's development. An emotionally intelligent AI could bring a lot of positive changes to the tech world, but it also poses some risks. Since emotional cues are so nuanced, it would take a long time for AI machines to perfect reading

them, and could potentially make big errors while in the learning stage. And some worry that an AI capable of responding to both emotional and situational signals could lead to the automation of more jobs.

- Theory of Mind in Artificial Intelligence refers to the capacity of machines to sense and comprehend human feelings. The concept is borrowed from psychology, which describes the human ability to feel others' emotions and predict others' actions based on such knowledge. Although we are far from achieving this type of understanding in AI, it has a huge potential as a technological revolution breakthrough. An emotionally intelligent AI can potentially drive a range of positive innovations, but it also poses certain risks. Due to the complexity of emotional cues, long periods are likely to pass before AI systems learn to dominate such cues with errors potentially arising during the process. Further, such an AI system has the potential to accelerate the automation of more and more jobs [116].
- Self-awareness Artificial Intelligence: Self-awareness AI refers to the hypothetical stage of AI where machines possess self-awareness. Often referred to as the AI point of singularity, self-awareness AI represents a stage beyond theory of mind and is one of the ultimate goals in AI development. It's thought that once self-awareness AI is reached, AI machines will be beyond our control, because they'll not only be able to sense the feelings of others, but will have a sense of self as well [132].

Furthermore, the type of intelligence needed to make decisions combines applied data science, social science, decision theory, and managerial science to optimize decision-making. To attain perceptual, cognitive, and decision-making intelligence, an essential infrastructure consisting of data, storage, computing power, machine learning (ML) algorithms, and AI frameworks, is required. AI applications are supported by learning the internal laws of data through training models. The application layer of AI has grown extensively, integrating deeply into fundamental areas of knowledge, significantly impacting our work and lifestyle [133].

3. Simulation Overview

Precision and efficiency are the prime goals in today's rapidly evolving industries, such as desalination. Engineers, designers, operators, investors, and decision-makers across industries must ensure hassle-free running of their systems, be it existing systems or during pre-construction phases before they construct them. Focus on operational integrity is paramount so that they get the best out of performance and sustainability in such complex environments.

This simulation is a replica model that replicates the operation of either current or conceptual systems, providing empirical proof for better-informed decision-making with the ability to analyze various alternative scenarios or variations within processes. In addition, incorporating virtual reality technologies can also be added to this experience to provide more immersive and interactive space to investigate and evaluate [134].

Simulation of systems is a critical engineering methodology that enables prediction of performance, optimization of design, and identification of problems that may arise even before physical prototypes are created. It is particularly valuable in cases where the actual system is inaccessible, unsafe, or in the conceptual or theoretical phase of development.

Moreover, system simulation might be described as the usage of computer-supported software packages to simulate and analyze sophisticated systems in several fields of discipline, for example, mechanical, electrical, thermal, fluid, and multi-physics. With the use of mathematical representations—like finite element analysis (FEA) in the case of structural simulations and multi-body dynamics in the case of mechanical system engineers can simulate reality. This provides the possibility to clearly see interactions within the parts in the system and thus ease decision-making processes while designing and debugging [135].

3.1. Simulation Types

Simulation can be broken down into three main types, as follows:

- Discrete Event Simulation: Modelling a system as it progresses through time, such as, factory operations (stamping, turning, milling) and traffic analysis (roads, networks, queues) [136].

- Dynamic Simulation: Contrary to discrete event simulation, dynamic simulation is directed towards the representation of systems as they transition through spatial parameters. This form of simulation is routinely utilized in disciplines like machine kinematics, human factors and ergonomics, aerodynamic testing, and virtual prototyping because it allows one to visualize the interaction between systems and parts within a three-dimensional setting [137].
- Process Simulation: This category deals with the simulation of physical interactions among two or more systems. It finds applications in several fields, such as in-service product modeling, product evaluations during production, and weather forecasting, thus helping in the understanding of the complexities involved in system interactions over time and space. By categorizing simulations in this manner, professionals and researchers can select the appropriate modeling style aptly to tackle specific problems or circumstances [138].

3.2. Engineers' Simulation Needs

In an industry such as desalination, which contains multiple processes and interconnected devices, Cyber-Physical Systems play a vital role. which integrate physical components, such as sensors, actuators, mechanical, electric, and chemical systems, with digital control through embedded software. Desalination stands as a prime example of an industry that seamlessly integrates all these components.

Engineers use multi-physics system simulation software to ensure that a system like desalination operates reliably, efficiently, and sustainably in real-world operations. Engineers use multi-physics system simulation software to model the behavior of each component and their interactions within the system as a whole. Using simulation to make virtual prototypes (digital models of the physical system) enables engineers to test and improve the desalination system safely in a virtual setting before building any physical parts. This approach is key to identifying design flaws early, reducing costs, and ensuring seamless interaction between hardware and software [139].

3.3. Simulation Advantages

There are a range of advantages to be gained through the use of simulation, including:

- Less Financial Risk: Simulation is less expensive than real-life experimentation. The potential costs of testing theories of real-world systems can include those associated with changing to an untested process, hiring staff or even buying new equipment. Simulation allows you to test theories and avoid costly mistakes in real-life [140].
- Exact Repeated Testing: A simulation model that enables users to investigate hypotheses and innovations time after time against the exact same circumstances. Therefore, this means testing and comparing different ideas under the same conditions.
- Examine Long-Term Impacts: A simulation cable of accurate modeling and predicting the impact of years of circumstances. This lets users illustrate both short and long-term impacts, which help decision investigation [141].
- Gain Knowledge to Improve Processes: Simulation advantages, no need to wait until the end of a project, but rather, the whole timespan of a process can be improved continually by comparing and testing hypotheses [142].
- Assess Random Events: Simulations can be particularly utilized to analyze random events, for instance, random staff absenteeism or disruptions in the supply chain [142].
- Test Non-Standard Distributions: A simulation can accommodate variable and non-standard distributions, rather than being restricted to fixed parameters. For instance, while modeling a supermarket, one may enter several consumer kinds who would traverse the store at differing velocities. A youthful entrepreneur acquiring a sandwich will navigate the shop differently than an elderly couple or a mother conducting a weekly grocery shopping with two youngsters accompanying her. Incorporating these variable characteristics enables a simulation to more precisely replicate reality [143].
- Encourages In-Depth Thinking: The activity of creating a simulation and inserting different parameters can cultivate problem-solving through the promotion of greater analytic consideration of a particular process or procedure. Considered reflection on these variables may

provide novel solutions, which can be different from ultimate outcome of the simulation [144].

- Increase Stakeholder Buy-In: A visual simulation also increases buy-in by associates, partners and stakeholders. It predicts results of any process improvement and how they were achieved, improving interaction with interested parties or even allowing a simulation-based sales presentation [145].

4. Machine Learning Overview

Machine learning theory involves principles that are crucial in successful and efficient learning in AI. It deals with the development of models capable of learning from experience by making predictions based on the interdependent relationships among variables. This has been the cause of considerable improvement in image and speech recognition [146]. It is also considered a subcategory of AI, enabling machines to learn from data through algorithms and to make predictions. These can be developed either by supervised learning, whereby algorithms detect patterns from labeled examples, or by unsupervised learning, whereby the algorithms search for more general patterns in unlabeled data [147]. it has grown as one of the major imperatives across industries and sectors. This is important because it offers insights into customer behavior trends and operational patterns of organizations, hence helping them in innovating product development [148].

Machine learning heavily relies on data, which provides the backbone while training and in the evaluation of models. Data consists of inputs, most commonly referred to as features, and outputs, usually labeled as labels. While training, the model learns patterns and tests new, unseen data against performance and the capability of generalization [149]. There are three basic techniques in machine learning: classification, clustering, and collaboration.

Classification involves assigning new observations to categories or labels using a training dataset that has already been classified [150]. The machine learning lifecycle involves defining a problem to solve, gathering data, cleaning and preprocessing data, exploratory data analysis, choosing a model, training a model, and testing its performance.

4.1. Types of Machine Learning

Machine learning algorithms are broadly divided into four types – supervised learning, unsupervised learning, semi-supervised learning, and reinforcement learning.

- Supervised Learning: In the case of supervised learning, machines learn from examples. The operator feeds the machine a well-defined dataset containing both the input and the corresponding output, therefore allowing the learning algorithm to learn how to relate them. While in supervised learning, the operator knows the right answers; the algorithm makes matches, gains insight from what it sees, and predicts. The operator would then assess the algorithm's predictions and make corrections; this goes on iteratively until the algorithm achieves a high degree of accuracy.
- Unsupervised Learning: The machine learning program analyzes data for trends without any response key or human input that would dictate a direction. Rather, it's on its own to study the data to find correlations and connections. Through unsupervised learning, the machine learning algorithm processes large datasets and attempts to make sense of the structure inherent in them. This could be in the form of clustering the data or arranging it in a more organized fashion. The newer data it gets to review, the more enhanced its ability to identify patterns and relationships becomes. Consequently, this iterative process allows the algorithm to improve its performance over time, leading to more accurate insights. The system is capable of making enhanced and informed decisions.
- Semi-Supervised Learning: It is the same as supervised learning except that it employs both labeled and unlabeled data. Tags indicate the output of labeled data, while unlabeled data do not. This combination allows the model to learn from both the explicit examples and the underlying patterns within the data. contain semantic tags so that the algorithm can comprehend them, while unlabeled data contain no such tags. Machine learning systems, employing both of them, learn to tag the unlabeled data.
- Reinforcement Learning: Reinforcement learning pertains to structured learning methodologies wherein a machine learning algorithm is provided with a defined set of actions, variables, and

terminal outcomes to adhere to. Following the established guidelines, the algorithm endeavors to investigate many alternatives and possibilities, evaluating each output to ascertain the optimal choice. Reinforcement learning guides the machine by trial and error. It assimilates prior experiences and then modifies its strategy to get the optimal outcome [149, 151-153].

4.2. Artificial Intelligence and Machine Learning Connection

Although AI and ML are two different concepts, they are interconnected. For a better understanding of how they relate to one another, consider AI as the big idea: to let machines or systems become perceptive, reasoning, acting, or adapting somewhat as humans do. Conversely, machine learning is a subset of AI that involves the study of enabling machines to learn independently from data to draw insights.

One good analogy is thinking of AI as an umbrella wherein all these various techniques and algorithms are embraced, with the most important being machine learning. Deep learning, robotics, expert systems, and natural language processing all fall into subfields that constitute this broad area. While AI creates the machinery to work in human-like intelligent modes, ML in general allows them to learn, through patterns, how to accomplish certain tasks without any particular type of programming [154].

4.3. Benefits of Using Artificial Intelligence and Machine Learning Together

AI and ML benefit organizations of all shapes and sizes, with new possibilities constantly emerging. In particular, as the amount of data grows in size and complexity, automated and intelligent systems are becoming vital to helping companies automate tasks, unlock value, and generate actionable insights to achieve better outcomes. Here are some of the business benefits of using AI and machine learning:

- Wider data ranges: Analyzing and activating a wider range of unstructured and structured data sources [155].

- Faster decision-making: Better data integrity, quicker data processing, and the reduction of human errors collectively render decision-making processes better informed and more timely [156].
- Operational efficiency: Enhancing operational efficiency while simultaneously decreasing expenses represents a vital goal.
- Integration of analytics: Integration of predictive analytics and insights into business reporting frameworks and applications is crucial to enable employee empowerment and informed decision-making within organizations [155].

4.4. Simulation and Machine Learning

A simulation relies on established rules, whereas machine learning identifies and establishes the fundamental rules that characterize a certain environment. It pertains to scenarios that are poorly described, for which no formal mathematical framework exists.

Experts develop a model in a simulation to assess the probability of diverse variables inside a complicated system consisting of several interacting components. Such systems can be seen as a "system of systems," shown by predicting market behavior influenced by several individuals and external factors. The "what-if" situations in simulations are integrated into the model.

Conversely, machine learning develops models using data sets with known outcomes. Upon constructing a model, predictions may be made by applying acquired patterns to novel datasets [157].

Simulation and machine learning are interconnected as both rely on models; nonetheless, they exhibit distinct differences. In simulation, the model is often understood as it adheres to the predetermined rules established for it.

In other words, we understand how to receive input information, do calculations, and ascertain the output according to a predefined set of rules that we have established. Nonetheless, the characteristics of the inputs for the simulation remain unidentified. Some of the inputs are stochastic variables, and their exact values are unknown. Historical data is utilized to calibrate a probability distribution for the input, or a probability distribution is derived via expert assessments. The objective is to ascertain a spectrum of results by randomly picking input values and repeatedly computing the output [158].

5. Artificial Intelligence, and Simulation Integration

Since the mid-20th century, engineers and scientists have used simulation to validate, test, and confirm their designs. Every model generated synthetic data by running simulation software, which entailed millions of calculations to know what works and what doesn't. Today, AI is combining these insights with real-time data, eliminating gaps in what's achievable, and accelerating simulation more cost-effectively than ever. Figure 32 shows the practical benefits of unifying AI with simulation in engineering [159].

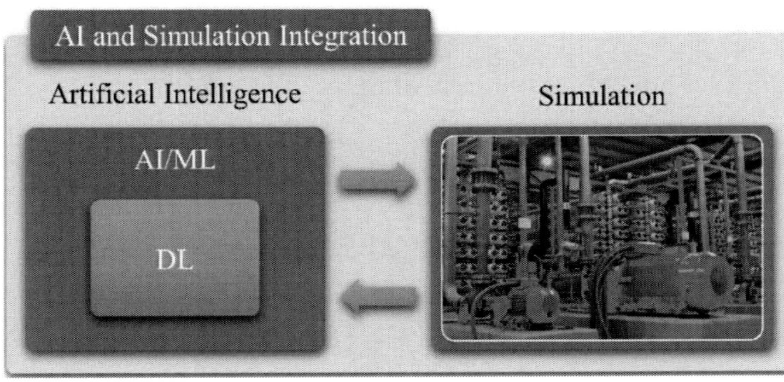

Figure 32. Artificial Intelligence Combined with Simulation Promises Real Benefits for Engineering.

AI, ML, and simulation are transforming human information processing and predictive capabilities, elucidating how the convergence of these advanced disciplines facilitates improved and expedited decision-making.

The relationship between AI, machine learning, and simulation: AI/Machine Learning and simulation are separate technologies that, when combined, enhance each other significantly. Initially, AI develops systems that replicate human behavior according to established norms. Machine Learning is a subfield of AI that allows computers to autonomously acquire rules from training data. Simulation examines real-world goods and processes using virtual models.

Although they are distinct ideas, they perpetually influence one another. Simulations employ AI and machine learning approaches to execute models more efficiently and effectively. Consequently, AI and ML employ simulation to generate synthetic data across many sectors and applications. Figure 33 illustrates the relationship among AI, ML, and simulation [160].

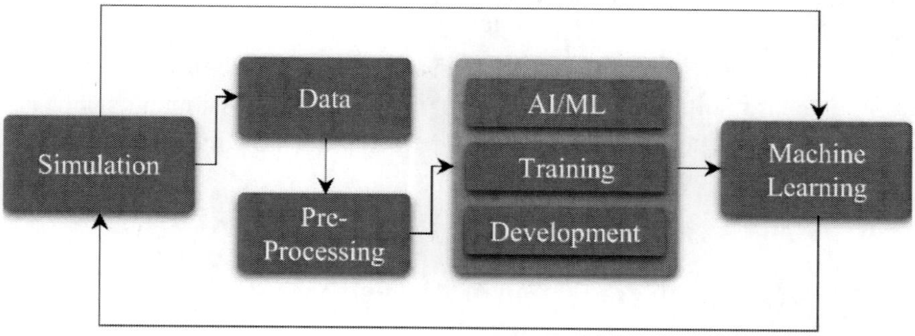

Figure 33. Relation Between AI, ML, and Simulation.

The impact of ML and AI has significantly impacted the growth of simulation technologies, where enhancing decision-making through speeding up and improving the quality of decisions is one of the key aims. AI and ML are crucial in making it happen. Using AI and ML, designers can start the process with a general specification, generate different design possibilities automatically, and speedily recognize the most potential ones. Several approaches, such as bottom-up and top-down methodologies, reduced-order models, and large language models (LLMs), enable our solvers to perform forecast calculations faster and more efficiently. LLMs such as ChatGPT can make the steps even more user-friendly. For instance, when an engineer is designing an electric vehicle, there might be a need to decipher lengthy specifications for the various parts, such as the battery, motor, and charger.

This would be time-consuming and may require expertise in certain areas. But with LLMs, all one has to do is just tell in simple words what one wants to be done, like, "Design an electric vehicle with a 2,000-mile range that goes from 0 to 60 mph in one second." The simulation then produces many possible solutions, some of which humans might not have thought of. AI also improves coding standards. In cases where programmers have the responsibility to update software, AI is able to recommend the most appropriate code snippet to repair particular problems, thereby improving productivity and customer service. By examining data that is received from numerous customer service interactions, virtual agents are in a position to provide effective first-level support, enabling experts to concentrate on more complicated inquiries. In the interpretation of a car specification, a human designer utilizes his or her brainpower and expertise. Conversely, AI can draw on a wide range of sources, such as questionable information or proprietary competitor data. This emphasizes the need for human intervention and methods such as explicable

AI. To ensure accuracy and assurance as AI and machine learning systems ingest large quantities of data, it is essential to use our own intellectual property for each system or process. Using the ChatGPT engine along with a strong database built from extensive experience and discussions in scientific and engineering fields can confirm the reliability of these simulations [161-163].

5.1. Role of Data and Simulation in AI Engineering

Information is the backbone of AI engineering. As earlier stated, AI is essentially based on mathematical concepts such as algorithms and calculations. This is the back-end training that no one sees or notices. Just like human teams need to be trained. For humans, this may be in the form of interactive training modules or practical sessions with seniors. However, AI training is different and still uses an information-based approach. Providing AI with the right data is a top priority for engineers, just as humans need accurate information to learn. In this respect, the integration of AI and machine learning can be seen not as a revolution but as a series of evolutions, just as educating an engineer or designer through a new design strategy or workflow yields a better outcome.

The key difference is that AI needs a huge amount of data to train properly since it lacks human intuition and the ability to make quick decisions. This makes many skeptical, but simulations can greatly improve predictive accuracy, process huge volumes of data, and provide real raw analytical power with speed. This is why integrating AI and machine learning with simulation is necessary for holistic AI engineering. Nowadays, advanced simulation software covers everything from FEA to CFD and electromagnetics, while countless varieties can be applied to virtually any engineering project. These tools enable engineers to capture the most accurate data and visualize how it performs. it also takes some of the mathematical burdens off engineers and designers by automating many calculations and coding techniques. Once the data required is collected, the team can switch back to simulation to proceed to the next step: training AI. An effective method of training is through the use of digital twin technology. In a digital twin, machine learning algorithms are used to create massive datasets that help speed up AI training while being less expensive than traditional physical methods of data creation. Figure 34 illustrates how AI operates.

AI plays a crucial part in creating hybrid digital twins. Another advantage of AI/ML-based methods is their capability to utilize historical data effectively. When first collected, information may need to be carefully sorted into primary, secondary, and disposable categories.

Less immediately valuable information contemplated gets stored on older media on storage drives and effectively becomes inaccessible. AI/ML can take this backlog, however, and turn what may appear to be outdated data into assets. Because AI learning takes enormous amounts of data (and mathematics), such dormant datasets can be reused and put to a new purpose. In that sense, AI/ML is comparable to unlocking the experience of experienced designers, which is a huge benefit. This also supports sustainability objectives because holding enormous amounts of data can result in wasteful energy expenditure, particularly if the data is not being utilized actively. Using AI/ML to filter through, amalgamate, and leverage prior data helps reduce energy consumption in the long term. To explore this aim, we must attempt to develop a data-rich development platform using an open framework upon which high-end machine algorithms can continuously draw and harness data from a myriad of sources.

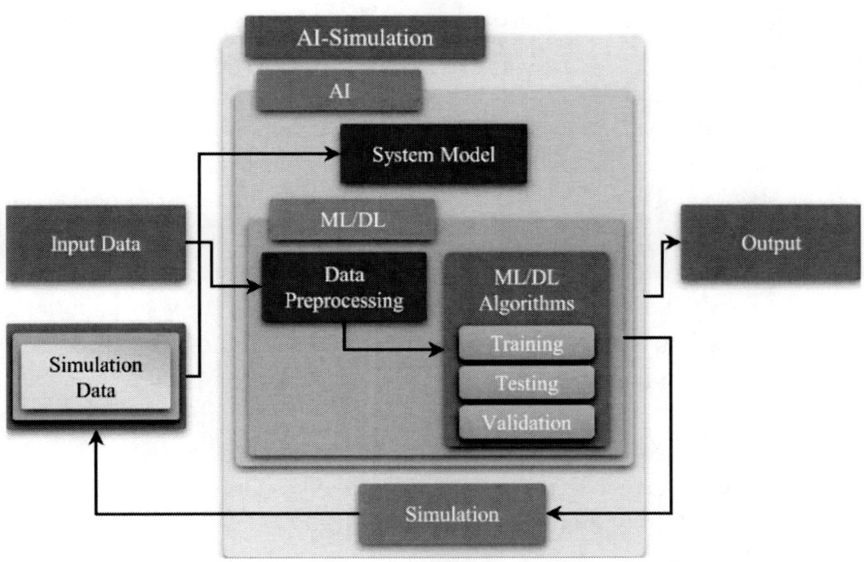

Figure 34. Concept of Combining Simulation and Artificial Intelligence.

5.2. Benefits of Combining AI and Simulation

The relationship between AI and simulation is characterized by symbiotic benefits where AI advances provide a great boost to the efficiency of simulation techniques. AI processes can accelerate simulation tasks up to 100 times, a scenario that is largely realized in AI-powered simulation systems as well as ML-based systems. For instance, machine learning is able to automatically create product designs through training neural networks that predict, optimize, and generate production-ready designs from historical data. Furthermore, machine learning is capable of enhancing the simulation process by tuning model parameters automatically and employing machine learning-based flow solvers.

Also, the integration of AI and ML can enhance design processes through topology optimization, streamline processes, and thus lead to cost-effectiveness. Despite the many advantages provided by simulation, AI/ML, and automation technologies, the role of human professionals remains essential. Engineers, designers, and other subject matter experts are crucial to train AI systems and enable the convergence of AI, simulation, and engineering practices. Their expertise is required to provide required training, knowledge, and guidance. Thus, these professionals are more capable of utilizing advanced AI-powered design software, which can greatly improve creative productivity in the design process [133, 164, 165].

Design and development processes were constrained by the time and precision of engineers manually performing simulations. The development of models for complicated systems consumed much time and professional expertise, thus hampering progress. Modern AI-driven simulations, on the other hand, make design and optimization processes more efficient in many fields, especially those where accuracy and effectiveness are very important, like materials science, aerospace, automotive, and electronics [154, 166-168].

AI-enhanced simulations are:

- Improved Efficiency: AI can examine past simulation history in a bid to rapidly detect complex patterns, as well as add new data to assess relationships within the dataset [169].
- Enhanced Accessibility: AI has the potential to democratize the use of simulations by bringing them within easy access for users without specialized expertise, achieved through intuitive web-based interfaces [170].

- More comprehensive: AI simulations enable the integration of multiple models to provide thorough representations of complex systems [171].
- Continually improving: The iterative engineering process enabled by AI and simulation enables engineers to improve their designs with fewer constraints [172].

5.3. Training AI Enables Fast Predictions

For AI to work effectively, it is essential that it demonstrate intelligence. Simulation of data can be one way to train AI in a certain domain. By applying past simulation data, AI programs can be trained in chosen domains of interest [173]. For example, in training AI in the domain of integrated circuits, the performance from simulations of circuit boards would be implemented in the software. In 3D physics, generative AI uses results from physics-based solvers to improve AI models and make predictions more quickly. One advantage of data-driven methods over the more common reduced-order modeling methods is that engineers don't have to set the parameters for the geometries to make the AI model.

This allows for performance predictions during design variations even when the geometric framework is altered. During the simulation phase, several geometric design alternatives are offered to the computer system to facilitate quick predictions of physical behavior. This optimizes the efficiency and convenience of design iteration, exploration, and optimization processes for a wide range of engineering users, including designers, systems engineers, and methods and tools experts. Additionally, conventional physics-based solving methods can be applied to validate the optimal design alternatives through large-scale simulations [174, 175].

5.4. The Future of AI/ML and Simulation

The mutually beneficial relationship between AI/ML and simulation will continue to increase efficiency for engineers and designers. As the technologies continue to pick up pace in multiple industries and applications, their use outside of engineering will also bring further advancements to humans.

As more people tap into the predictive powers of these potent technologies, possibilities for innovation will increase exponentially. Most simulation software companies now use machine learning platforms that allow engineers to specify complex product requirements and rapidly investigate and predict the performance of new ideas through various stages of design. Thousands of new ideas that would not even occur to a human designer can be generated by one. Simulation can check all these designs very rapidly, and designers can choose the best one. The intersection of simulation and AI/machine learning has many opportunities. As indicated in Figure 35, past data can be used to predict new design performance [176, 177].

6. Artificial Intelligence and Simulation Applications for Desalination

AI technologies possess the capacity to transform the operational efficiency, performance forecasting, and decision-making capacities of desalination facilities. Utilizing diverse AI methodologies, including Machine Learning, neural networks, and predictive analytics, we can enhance the operational parameters of desalination systems, such as energy consumption, water production, and maintenance scheduling, resulting in a more efficient and sustainable future for desalination.

AI-driven predictive models may be created to anticipate the performance of desalination facilities, encompassing water quality and process constraints. Thus, facilitating proactive maintenance and improved operational planning. Furthermore, the incorporation of AI-driven decision support systems can substantially improve the decision-making processes of desalination plant operators. This enables them to make more informed, data-driven decisions concerning resource allocation, process modifications, and overall plant management. Case studies and empirical examples have illustrated the effective application of AI technology in desalination facilities, emphasizing the concrete advantages and insights gained. Nonetheless, obstacles, best practices, and future trends must be considered regarding the application of AI to optimize and enhance desalination plant operations and performance.

Artificial Intelligence and Machine Learning have been integrated into several engineering disciplines, including purification and water management applications. They are essential in managing the inevitable fluctuations in process conditions inside desalination facilities, thereby optimizing operational efficiency. AI modeling capabilities are very advantageous in water filtration and sewage treatment techniques, since they automate these processes, rendering them more efficient, cost-effective, and substantially minimizing the likelihood of human mistakes. Figure 36 illustrates the parameters of desalination systems utilizing AI-ML [109, 110].

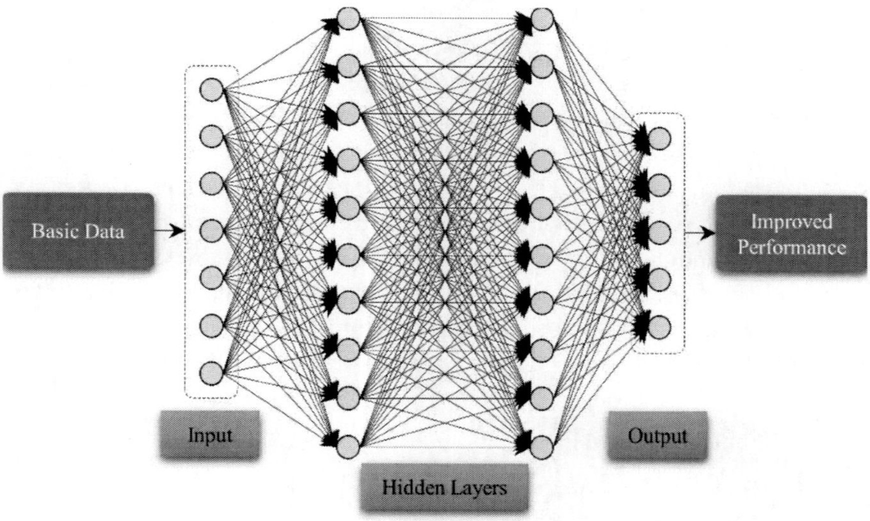

Figure 35. Leverages Past Data to Predict the Performance of New Designs.

6.1. Desalination AI Engineering Application Overview

Recent progress in artificial intelligence has made it much easier to use technologies that desalinate water. AI systems use a variety of methods, such as machine learning, deep learning, natural language processing, and computer vision, to understand large datasets and make smart decisions or predictions.

These high-tech systems use artificial intelligence to make decisions that are more effective and better. They do this by using a lot of data and machine learning to analyze information, guess what will happen, and boost operational productivity. AI can play a big role in desalination in several important areas, such as, economic efficiency, water quality improvement, site selection, and energy management.

Economic Efficiency: Through leveraging the interconnected benefits of desalination alongside power and chemical manufacturing during both the design and execution phases. Water Quality Improvement: Through in-depth data analysis to ensure better water quality and maintain membrane integrity. Site Selection: Based on geographical and environmental data to identify the most suitable sites for desalination plants. Energy Management: Reducing energy consumption in the desalination process.

Practical uses verify the advantages of artificial intelligence to the desalination industry. One important one is predictive maintenance, which uses machine learning to figure out what is most likely to go wrong with machinery before it does. This lets maintenance happen before it breaks down, with little to no interference. Also, AI-based technologies make it easier to keep an eye on and manage operations, so that when something goes wrong, corrective actions can be taken right away to make things run more smoothly.

The use of AI-driven optimization algorithms is central to minimizing energy consumption and operational expenditure. Using these AI techniques successfully shows that they have a lot of benefits and that AI technologies should be used more in the water desalination industry. This will improve efficiency, cost-effectiveness, and overall performance [109].

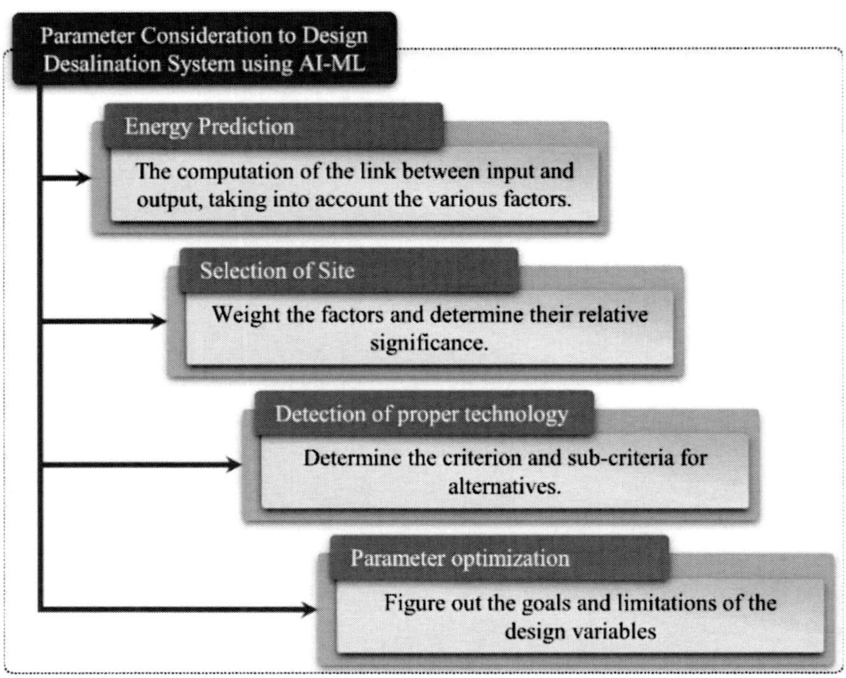

Figure 36. Areas in Desalination Systems where AI-ML is Applied.

6.2. Importance of AI Engineering in Desalination

Artificial intelligence is profoundly transforming the desalination industry by introducing significant operational improvements. AI is being utilized across various operational dimensions, enhancing the reliability of desalination plants. As said before, in predictive maintenance, AI algorithms can analyze sensor data and operational patterns to predict when equipment may require maintenance, allowing for proactive repairs before failures occur.

This helps improve the reliability of desalination plants and decrease the downtime. For anomaly detection, AI models can identify unusual operational conditions or patterns that may indicate equipment issues or process inefficiencies. This enables prompt interventions to address problems before they escalate.

AI-powered systems can also continuously monitor and optimize various parameters, such as energy consumption, feed water quality, and product water output. This performance optimization further enhances the reliability and overall efficiency of desalination plants, making them more cost-effective. AI enhances the reliability of desalination plants by predicting failures before they occur, leading to greater operational efficiency and cost reduction.

Deep learning provides sophisticated analytical capabilities that can interpret complex data patterns and accurately predict outcomes. It enables enhanced predictive analytics for equipment failures and early faults. detection by finding small changes in operational data like voltages or currents; and process optimization by learning all the time and making changes to system parameters to get the most out of operational data. AI systems in desalination plants use sensors and IoT devices to continuously monitor various parameters.

AI algorithms analyze this data to ensure membrane integrity, detect leaks, and monitor overall water quality. This real-time monitoring helps maintain optimal conditions throughout the desalination process. In this continuous analysis of operational processes, optimization of algorithms generates the most efficient settings for the plant's machinery and processes. These algorithms employ sophisticated mathematical models to achieve balance and efficiency. adjust critical parameters such as energy consumption, chemical dosages, and pressure levels.

By doing so, they aim to achieve optimal water recovery rates while minimizing energy use and operational costs. It enhances the plant's sustainability and ensures that it operates within the most cost-effective parameters. Additionally, these algorithms can adapt to changes in input water quality or environmental conditions, ensuring consistent output quality and operational efficiency.

Conclusion

The use of AI in the desalination industry marks a significant improvement in water treatment efficiency and sustainability. AI enhances process optimization, operational reliability, and data-driven decision-making. AI technologies improve desalination by optimizing energy use, enhancing water quality, and lowering costs. They enable predictive analytics and real-time monitoring, which streamline operations and reduce waste. Predictive maintenance strategies help anticipate equipment failures, minimizing downtime and extending equipment lifespan. AI also allows desalination operators to make informed decisions by analyzing complex data and identifying patterns that may be missed by humans. This capability supports better planning and management of desalination systems.

In terms of sustainability, AI aids in integrating renewable energy sources and reducing the environmental impact of desalination, aligning with global sustainability goals. However, implementing AI in desalination comes with challenges, such as ensuring data quality, meeting regulatory standards, and protecting against cybersecurity threats. Organizations need strong data governance and compliance practices to address these issues. Successful AI integration requires collaboration among AI specialists, engineers, and operational staff, along with ongoing training to adapt to new technologies. A strategic approach, starting with pilot projects and scaling gradually, is essential for managing risks and effectively applying AI.

In summary, AI has the potential to transform the desalination industry by driving innovation, improving efficiency, and promoting sustainability. A strategic and collaborative approach to AI integration can lead to enhanced operational excellence and a more sustainable future in water resource management. Embracing learning and adaptation is key to realizing AI's full potential.

Chapter 6

Filtration Unit Modeling, Simulation and Validation

Abstract

This chapter introduces a mathematical model formulated to examine the functioning of a filtration unit utilized in water treatment, specifically highlighting media filtration concepts. The four principal parts of the filtration unit are the booster pump, multi-media filter (MMF), activated carbon filter (ACF), and cartridge filter (CF). The model utilizes characteristic equations to explain the functioning of every component of the system, culminating in an explanation of the overall filtration process. The chapter formulates separate mathematical equations for each of the components and addresses the simulation program that integrates the equations into a comprehensive model of a filtration system. Results seek to maximize design and operation of filtration systems for effective treatment of feed water in RO operations.

Keywords: mathematical modeling, desalination system, filtration units, pretreatment, hydraulic accumulator, booster pump, cartridge filter, carbon filter, multi-media filter

1. Introduction

Mathematical modeling definition is the application of mathematical calculations and algorithms to predict the working and performance of functions. This process involves articulating real-world problems through mathematical equations and using different methods to find solutions.

A mathematical model simulates the behavior of the filtration unit under various operating conditions, which helps predict performance indicators including water quality and pressure drop [178]. Although mathematical equations cannot fully capture all aspects of real-world problems, their solutions can often be more pragmatic due to the simplifying assumptions they incorporate. Filtration unit mathematical models are primarily based on mass balance and energy balance principles [179]. These mathematical models are essential for optimizing both the design and operational aspects of the filtration unit by evaluating the circumstances of varying in conditions on operation environment, such as water flow and deferential pressure. Additionally, these modeling equations can help select appropriate media and configurations, assess the influence of filtration performance, and estimate system lifetime [180].

For water treatment researchers and engineers, filtration unit mathematical modeling is a crucial tool that improves the understanding and optimization of these systems. The process of finding the most effective choices among various options is called optimization, emphasizing the importance of developing an accurate mathematical representation. to ensure the filtration unit operates at its optimum performance, in addition to the probability of total cost reduction. Moreover, mathematical modeling and optimization play a vital role in the design process.

Advances in computer technology have led to impressive results, toward the creation of advanced software tools that facilitate meshing, geometric modeling, and presentation of results, as well as sophisticated analytical algorithms [181]. A mathematical model is developed to investigate the operation of filtration units working with the principle of media filtration. The model has been developed for filtration unit consisting of; booster pump, multi-media filter, carbon filter and cartridge filter.

The model is obtained by applying the characteristic equations describing the performance of the system components. This chapter presents the mathematical model, simulation and validation of the filtration unit components to obtain the mathematical model of the filtration unit as a whole. Every component will be explained along with their mathematical equations individually. The simulation program assembles the equations into a comprehensive model for the filtration system. Figure 37 shows parts of the pre-treatment unit and the Reverse Osmosis (RO) unit.

2. Filtration Unit Characteristics Mathematical Model

The filtration unit (pre-treatment unit) encompasses a centrifugal booster pump alongside a filtration group that comprises three distinct filter types: MMF, ACF, and CF. The mathematical presentation of each part is discussed in detail based on (Euler's head versus flowrate) equations in addition to the influence of hydraulic losses in the flow passages [182].

2.1. Booster Pump

The booster pumps are essential parts of the units designed to boost water pressure and flow in a hydraulic system. They act as an intermediary between the water source and several endpoint reverse osmosis systems. When they increase the water pressure, the flow dramatically increases, similar to a fan where the movement of the blades facilitates an increased flow [183].

The booster pump employed in this model works on the principle of a centrifugal mechanism, where its pump head depends on the rate of flow developed at a fixed rotational speed. The characteristic equation of the pump head, formulated from the energy equation, takes into account the few numbers of vanes with the hydraulic losses experienced within the flow channels [182] By adjusting of parameters of the pump, such as the impeller shape and speed, with some parameters being kept constant, the characteristic equation for a booster pump of adjusted impeller geometrical specifications, operating at a fixed speed, is presented in mathematical equation as shown in Equation (37).

$$H = k_1 - k_2 Q - k_3 Q^2 \tag{37}$$

where: H: Head, (m); Q: Flowrate, (m^3/s); and k_1, k_2, k_3 are coefficients of the pump, impeller geometry and operation speed (k_1 (m), k_2 (s/m^2), and k_3 (s^2/m^5)).

Characteristic equation of the centrifugal pump is developed from the pump real head using Equation (38).

$$H = H_t - h_{LF} - h_{LL} \tag{38}$$

where: H: Generated head, (m), H_t: Theoretical head, (m), h_{LF}: Friction losses in the pump passages, (m) and h_{LL}: Local losses (shock losses), (m). Generated head can be calculated using Equation (39) and (40). Friction losses and local losses using Equation (41) and (42).

$$H_t = \lambda \ H_{th\infty} \tag{39}$$

$$\lambda = 1 - \frac{\pi \ sin \beta_2}{Z_V} \tag{40}$$

$$h_{LF} = k_{LF} \times \ Q^2 \tag{41}$$

$$h_{LL} = k_{LL} \times (Q - Q_S)^2 \tag{42}$$

where: λ: *Van no. coefficient of pump impeller. (Experimentally formula),* $H_{th\infty}$: *Pump theoretical head, (m),* Q: *Flow rate,* (m^3/s) *and* Q_S: *Shock less flow rate (at which flow enters the impeller without shock),* (m^3/s).

Assume that, 100% volumetric efficiency, zero blade thickness, and radial inlet flow. So $H_{th\infty}$ will have the form shown in Equation (43).

$$H_{th\infty} = \frac{(\pi\ N\ D_2)^2}{3600\ g} - \frac{N}{60\ g\ b_2\ tan\ \beta_2} \times Q \tag{43}$$

where: β_2: Impeller exit blade angle, (deg.), D_2: Impeller exit diameter, (m), N: Pump speed, (r.p.m), g: Acceleration of gravity, (m/s^2) and b_2: Impeller exit width, (m).

Then the characteristics equation takes the form as shown in in Equation (44), (45), and (46).

$$H = \frac{\lambda\ (\pi\ N\ D_2)^2}{3600\ g} - \frac{\lambda\ N}{60\ g\ b_2\ tan\ \beta_2} \times Q - k_{LF}\ Q^2 - k_{LL}(Q - Q_S)^2 \tag{44}$$

$$H = \frac{\lambda(\pi\ N\ D_2)^2}{3600\ g} - \frac{\lambda\ N}{60\ g\ b_2\ tan\ \beta_2} \times Q - k_{LF}\ Q^2 - k_{LL}Q^2 + 2k_{LL}Q_SQ - k_{LL}Q_S^2 \tag{45}$$

$$H = \left[\frac{\lambda\ (\pi\ N\ D_2)^2}{3600\ g} - k_{LL}Q_S^2\right] - \left[\frac{\lambda\ N}{60\ g\ b_2\ tan\ \beta_2} - 2k_{LL}Q_S\right] \times Q - [k_{LF} + k_{LL}]\ \times Q^2 \tag{46}$$

By rearranging the equation, the characteristic equation of a centrifugal pump of given impeller geometry, working at constant speed, may be expressed in the form as shown in Equation (47).

$$H = k_1 - k_2Q - k_3Q^2 \tag{47}$$

where: k_1, k_2, k_3: Coefficients for a given inputs pump, impeller and speed and H: Head, (m). Which can be calculated using Equations (48), (49), and (50).

$$k_1 = \frac{\lambda\ (\pi\ N\ D_2)^2}{3600\ g} - k_{LL}Q_S^2 \tag{48}$$

$$k_2 = \frac{\lambda\ N}{60\ g\ b_2\ tan\ \beta_2} - 2k_{LL}Q_S \tag{49}$$

$$k_3 = k_{LF} + k_{LL} \tag{50}$$

The characteristic equation of the pump has been obtained from the curve fitting technique (at 2900 rpm) as shown in Equation (51).

$$H = 50.152 + 0.1698 \times Q - 0.0103 \times Q^2 \tag{51}$$

The positive sign in the second term in the pump characteristic equation is due to the shock loss term (at the entrance or outlet of the blade), which usually results in a peak head (surge point).

2.2. Filtration Group

In RO process of desalination, the quality of feed water must always be maintained to ensure proper operation. Poor feed water quality can decrease the lifespan of RO membranes, more frequent operational downtime, and increase maintenance costs. Therefore, it is essential to remove or reduce harmful substances contained in the raw water to acceptable levels. Such undesired water characteristics include TSS or total suspended solids, TOC representing dissolved organic pollutants, liquids such as oil and grease, and sparingly soluble metals and salts.

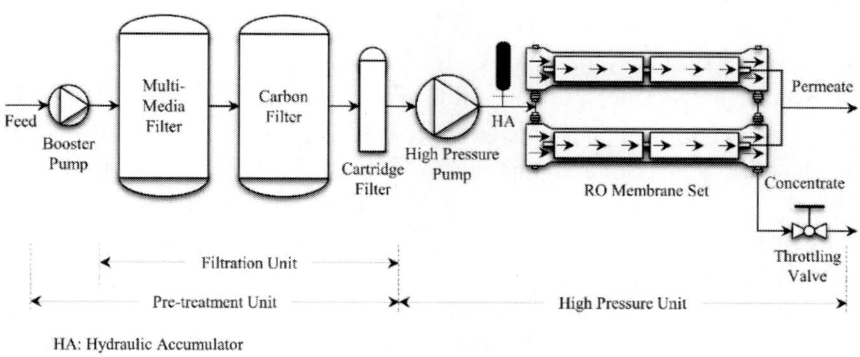

Figure 37. Components of Filtration and RO Units.

Accordingly, the pre-treatment stage is essential and becomes an indispensable part of the RO system [183]. Direct filtration techniques such as mono-media, dual-media, and/or multi-media, are the most common choices for the pretreatment stage of saline water before it is fed into RO system. This filtration process is mainly dependent on a series of intricate physical and chemical interactions, where adsorption is paramount. As the feed permeates through the filtration media, suspended particulates adsorbed to the surfaces of the media material [183, 184]. In order to simulate this process, the following forms represent the filtration group's calculations in detail for each type.

2.2.1. Multi-Media Filter Characteristics Equations

Media filtration is a well-known technique in desalination systems, which is usually incorporated for the removal of residual suspended solid contaminants from feed water, through a bed composed of porous granular materials. This process effectively removes suspended particulates as the water passes through a filtration medium composed of sand or gravel [185].

In the course of filtration, suspended particles stick to the surface of the discrete media grains and afterward are caught in the interstices of the medium. Granular media filters are in application in most operating full-scale seawater reverse osmosis (SWRO) plants [186]. The critical parameters that determine the selection of filtering media include type, homogeneity, particle size, and the depth of the material.

Filters can be classified into two kinds: mono-media and mixed-media, based on the number of designed layers. multiple granular media such as sand, gravel, diatomaceous earth, activated carbon, and anthracite, or fibrous media such as sponge, and cotton, have different design specifications which make them more efficient in filtration processes [187].

Granular media filters are helpful for particulate matter in the size range of 1 to 0.1 micrometers. However, A typical configuration of dual-media filters includes a layer of anthracite that is 0.4 to 0.8 meters deep placed over a layer of sand ranging from 1 to 2 meters in depth is advantageous in enhancing the removal of soluble organic matter from seawater by biofiltration processes; in such cases, the depth of the anthracite layer is typically increased to a depth of 1.5 to 1.8 meters [188].

The expressing equations of MMF can be presented as shown in Equation (52).

$$\Delta p_{mmf} = k_4 \times Q + k_5 \times Q^2 \quad (52)$$

where: Δp_{mmf}: pressure drop across the multi-media filter, (Pa); k_4: friction coefficient of multi-media filter porous media, ($N.s/m^5$); k_5: friction coefficient of multi-media filter vessel, ($N.s^2/m^8$).

2.2.2. Activated Carbon Filter Characteristics Equations

Activated carbon filtration is a physio-chemical technique which recognized as an advantageous option that relies on the adsorption basis, whereby dissolved organic compounds and heavy metals can be extracted from aqueous solutions [189]. activated carbon filtration is highly applied in various water treatment strategies because of the material's effectiveness in removing particulate matter, heavy metals and organic materials [190].

Activated carbon is generally classified into two different classes: powdered activated carbon (PAC) and granular activated carbon (GAC). The application of PAC is particularly advantageous in the removal of high amounts of organic and inorganic impurities, due to its large surface area and high degree of micro-porosity. Being the most commonly used absorbent, PAC is often used in combination with sand filter to enhance the overall efficiency of the treatment process and control of fouling also shows a remarkable enhancement in the removal of dissolved organic matter near [191].

The expressing equations of ACF can be presented as shown in Equation (53).

$$\Delta p_{cf} = k_6 \times Q + k_7 \times Q^2 \quad (53)$$

where: Δp_{cf}: pressure drop across the carbon filter, (Pa); k_6: friction coefficient of carbon filter porous media, ($N.s/m^5$); k_7: friction coefficient of carbon filter vessel, ($N.s^2/m^8$).

2.2.3. Cartridge Filter Characteristics Equations

Efficient design and operation of pretreatment systems are critical to ensuring effective removal of potential contaminants in the feedwater used in RO systems. Traditionally, RO based desalination plants incorporate low-micron range CF units as the last line of defense before high-pressuring pumps and the associated RO Membranes end. Usual CFs operate in dead-end flow mode, where all the incoming feed is permeated through the filter bed, thus trapping solids and particulate matter inside the filter medium. Cartridge filters have absolute and nominal ratings as the characteristic definitions.

Absolute rating refers to the diameter of the biggest sphere-shaped particle that can flow through the filter bed under designed test conditions, and therefore it refers to the diameter of opening pores in the filter. Nominal rating is defined as a filter's efficiency in stopping a minimum percentage of the removed or rejected solid, which is greater than the nominated micron size that is passing through the filter media, indicating its effectiveness in rejecting solid contaminants.

CFs are primarily designed to act as a guard against the entry of sand grains and other contaminants that might have passed through the upstream media filters. When the media filter performance cannot meet the pre-established requirements, the CF acts as a secondary filtration stage and also covers other ancillary activities such as removing suspended solids and controlling the silt density index (SDI) [192]. as shown in Equation (54) represent pressure drop through CF.

$$\Delta p_{carf} = k_8 \times Q + k_9 \times Q^2 \tag{54}$$

where: Δp_{carf}: pressure drop across the cartridge filter, (Pa); k_8: friction coefficient of cartridge filters element, ($N.s/m^5$); k_9: friction coefficient of cartridge filter vessel, ($N.s^2/m^8$).

The filtration group of the desalination plant is a depth filtration type. Characteristic equation of each filter is the summation of pressure drop in the filter vessel and pressure drop in the porous media layers.

2.2.4. Pressure Drop in The Filter Vessel

The pressure drop in the filter vessel can be developed by using the equation of head losses (friction and local losses) for ducts and pipes [182]. Friction losses can be calculated using Equation (55), while local losses can be calculated using Equation (56)

$$h_l = f \frac{L_f}{D_f} \frac{\overline{V}_f^2}{2} \tag{55}$$

where: h_l: Friction losses, (m), f: Friction losses factor, L_f: Filter height, (m), D_f: Filter diameter, (m) and V_f: Average flow velocity, (m/s).

$$h_{lm} = K_f \frac{\bar{V}_f^2}{2} \tag{56}$$

where: h_{lm}: Local losses, (m), K_f: Local losses coefficient and V_f: Average flow velocity, (m/s).

The total head losses in the filter vessel can be calculated using either Equation (57) or (58).

$$h_{fv} = h_l + h_{lm} \tag{57}$$

$$h_{fv} = f \frac{L_f}{D_f} \frac{\bar{V}_f^2}{2} + K_f \frac{\bar{V}_f^2}{2} \tag{58}$$

where: h_{fv}: Head losses in filter vessel, (m).

For specific filter diameter, length and fluid, pressure drop in the filter vessel can be expressed as shown in Equation (59).

$$\Delta p_{fv} = k_{fv} \times Q^2 \tag{59}$$

where: Δp_{fv}: Pressure drop in filter vessel, (bar) and k_{fv}: Coefficient for specific, filter diameter, height and fluid. k_{fv} can be calculated using Equation (60).

$$k_{fv} = \frac{L_f}{D_f} \frac{\rho g f}{2A^2} + \frac{\rho g}{2A^2} \tag{60}$$

where: A: Filter cross section area, (m^2), g: Acceleration of gravity, (m/s^2) and ρ: Density of fluid, (kg/m^3).

2.2.5. Pressure Drop in Porous Media Layers

Pressure drop in the porous media layers can be developed from *Ergun* equation because of Reynolds number (mixed flow i.e., both viscous and inertial forces are included). The Reynolds number for the used media in filtration unit is calculated using Equation (61).

$$R_{e_p} = \frac{\rho \bar{V}_o D_p}{\mu} \tag{61}$$

where: R_{ep}: Reynolds number for media particle, $\bar{V}_o$: Superficial velocity above bed, (m/s), μ: Absolute viscosity of fluid, (kg/m.s), ρ: Density of fluid, (kg/m^3) and D_p: Media partial equivalent diameter, (m).

Ergun equation for pressure drop in the porous media can be represented as described on Equation (62).

$$\Delta p_{pm} = 150\frac{L_{pm}\mu(1-\varepsilon)^2}{g_c\Phi_S^2 D_p^2\varepsilon^3}\bar{V}_o + 1.75\frac{L_{pm}\rho(1-\varepsilon)}{g_c\Phi_S D_p\varepsilon^3}\bar{V}_o^2 \tag{62}$$

where: Δp_{pm}: Pressure drop in porous media, (bar), L_{pm}: Total bed depth, (m), μ: Absolute viscosity of fluid, ε: Porosity and Φ_S: Sphericity.

For specific porous media, length of porous media, filter diameter and fluid; Pressure drop in the porous media layers can be expressed as shown in Equation (63).

$$\Delta p_{pm} = k_{pm1} \times Q^2 + k_{pm2} \times Q \tag{63}$$

where: $k_{pm1,2}$: Coefficients for specific porous media, length of porous media, filter diameter, and fluid.

These coefficients can be calculated using both Equation (64) and (65).

$$k_{pm1} = 1.75\frac{L_{pm}\rho(1-\varepsilon)}{g_c\Phi_S D_p\varepsilon^3 A^2} \tag{64}$$

$$k_{pm2} = 150\frac{L_{pm}\mu(1-\varepsilon)^2}{g_c\Phi_S^2 D_p^2\varepsilon^3 A} \tag{65}$$

The total pressure drop in the filter is the summation of pressure drop in the filter vessel plus pressure drop in the porous media layers as expressed on Equation (66).

$$\Delta p_{filter} = (k_{fv} + k_{pm1}) \times Q^2 + k_{pm2} \times Q \tag{66}$$

Equation (67) representing pressure drop through multi-media filter (rapid filtration, sand media), Equation (68) representing pressure drop through activated carbon filter (rapid filtration, activated carbon media), and Equation (69) representing pressure drop through cartridge filter (rapid filtration).

$$\Delta p_{mmf} = k_4 \times Q + k_5 \times Q^2 \tag{67}$$

where: Δp_{mmf}: pressure drop across the multi-media filter, (Pa); k_4: friction coefficient of multi-media filter porous media, ($N.s/m^5$); k_5: friction coefficient of multi-media filter vessel, ($N.s^2/m^8$).

$$\Delta p_{cf} = k_6 \times Q + k_7 \times Q^2 \tag{68}$$

where: Δp_{cf}: pressure drop across the carbon filter, (Pa); k_6: friction coefficient of carbon filter porous media, ($N.s/m^5$); k_7: friction coefficient of carbon filter vessel, ($N.s^2/m^8$).

$$\Delta p_{carf} = k_8 \times Q + k_9 \times Q^2 \tag{69}$$

where: Δp_{carf}: pressure drop across the cartridge filter, (Pa); k_8: *friction coefficient of cartridge filters element, ($N.s/m^5$);* k_9: friction coefficient of cartridge filter vessel, ($N.s^2/m^8$).

3. Static Characteristic Measurement

Special test rig has been developed for every single component of the filtration system, with measuring instruments suitably located to effectively evaluate the standing performance for each element under different testing adjustments. Experimental test rig is classified into two general categories: functional hydraulic circuit, which incorporates the component being tested, and measuring instruments that are allocated for the acquisition of data. The working arrangement consists of a storage tank, pressure-boosting pump, throttle valve, and the test object. Figure 38 illustrates the filtration system hydraulic equipment, together with the respective measurement instruments. The measuring instruments utilized in this evaluation; are a flow rate transducer, a pressure transducer, a tachometer, and a conductivity transducer. Before proceeding with taking the necessary measurements, it is essential to carry out calibration processes on the measuring devices. The data acquisition system involves data acquisition card, a microprocessor, and an industrial computer. The data acquisition card is an analog-to-digital (A/D) input device and is installed within a control panel that contains other input and output cards. The cards are all networked and operated by the microprocessor, which then communicates with the industrial computer. The transducers are interfaced to the A/D input card. Proper error analysis was carried out on all the data gathered for accuracy and reliability.

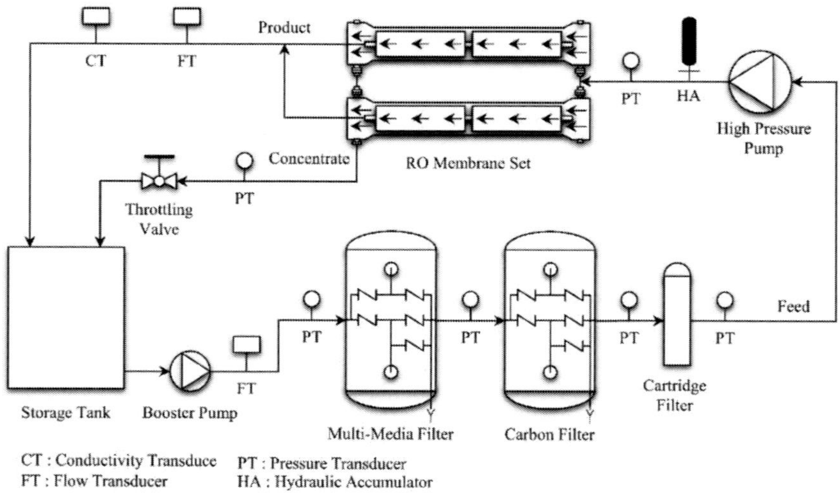

Figure 38. Hydraulic Circuit of RO Plant (Test Rig).

3.1. Booster Pump Test Rig

Hydraulic circuit used for testing the booster pump presented in Figure 39. The circuit includes a storage tank, a booster pump, a flowrate transducer, pressure transducers, and a tachometer. The experimental setup is a closed-loop system, starting from a storage tank that contains feed water. The feed flows from the storage tank through the piping system by the pressure generated by the booster pump. After that, feed flow back to the storage tank after being regulated by a flow control valve.

The operational parameters of the pump are measured by testing devices involving pressure transducers and a flow meter transducer, which are contacted into the suction and delivery lines of the pump. A tachometer is also fitted on the pump's electric motor shaft to note down the operating speed at each individual test point.

The influence of the speed variation is considered to improve the measured head and flow rate of the pump. The measured operational parameters of the test pump are suction pressure, delivery pressure, and electric motor speed, examined over a varying range of pump flow rates from

0 to 15 m³/hr. Figure 40 presents a comparative study of the booster pump characteristics as provided by the manufacturer, measurement results from experimentation, and calculated results from proposed empirical equations. The measurements attained the lowest pumping load, as estimated by the piping system of the hydraulic circuit.

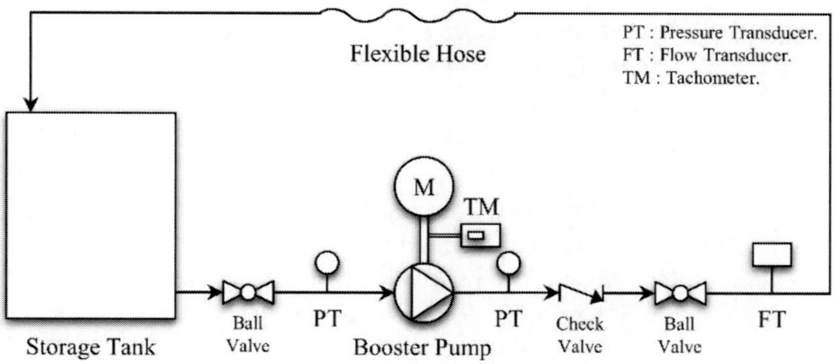

Figure 39. Hydraulic Circuit of Booster Pump (Test Rig).

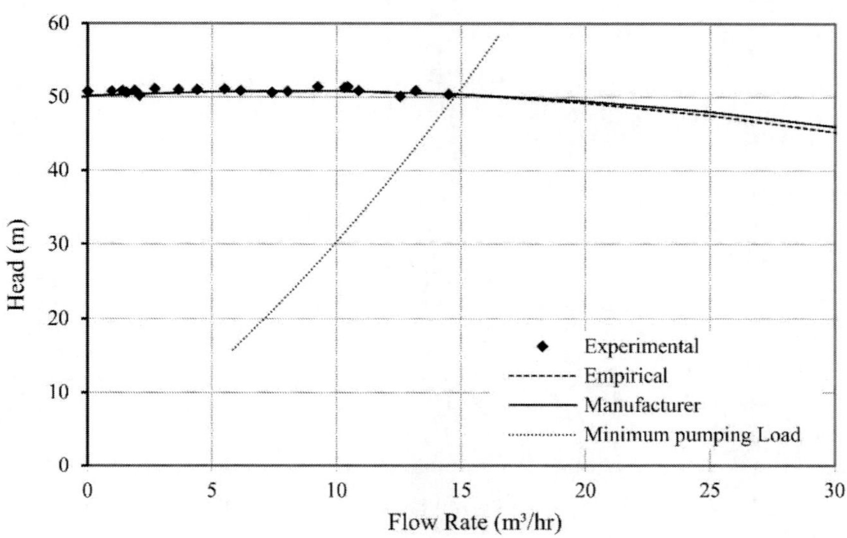

Figure 40. Comparative Analysis of Booster Pump Performance Provided by the Manufacture Sources, the Measured Performance, and Proposed the Empirical Formula.

The data presented in this figure shows a close agreement between the manufacturer's data, the experimental measurement results, and the suggested empirical equations within the range of operation.

3.2. Filtration Group Test Rig

Hydraulic cycle of the test rig used in the MMF, ACF, and CFs, displayed in Figure 41. The measuring devices used to confirm the characteristics are a flow meter transducer and two pressure transducers. The parameters to be measured are input pressure and output pressure for different flow rates between 5 and 15 m^3/hr.

The pressure drop-flowrate characteristics of components of the filtration unit were analyzed to derive their coefficients. From the results, it can be seen that pressure drop as derived from the curve fitting is polynomial, especially under high flow rates due to the filter body design. Within the range of operation, however, pressure drop is linearly related.

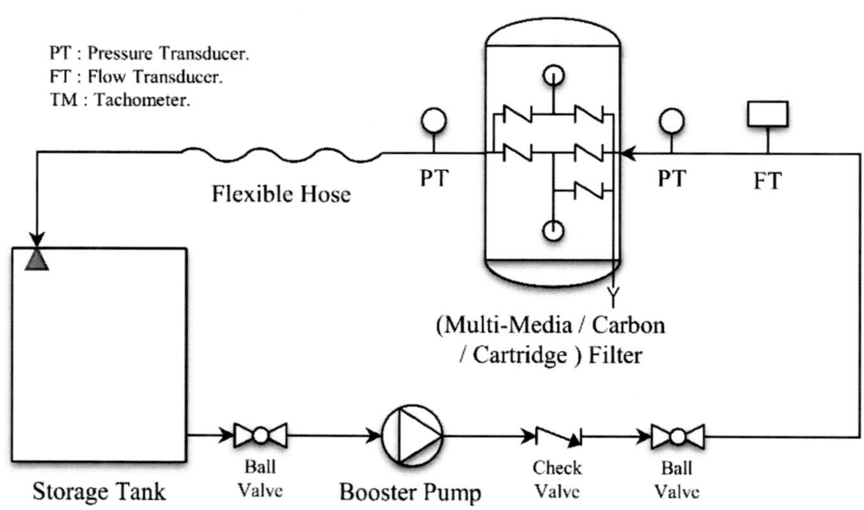

Figure 41. Hydraulic Circuit of Multi-Media, Activated Carbon, and Cartage Filter (Test Rig).

Equation (70), Equation (71), and Equation (72) representing curve fitted characteristic equation for MMF, ACF, and CF sequentially.

$$\Delta p_{mmf} = 0.005382 \times Q^2 - 0.009967 \times Q \quad (70)$$

where: Δp_{mmf}: pressure drop across the multi-media filter, (bar), Q: flow rate, (m^3/hr.).

$$\Delta p_{cf} = 0.003214 \times Q^2 - 0.01103 \times Q \quad (71)$$

where: Δp_{cf}: pressure drop across the carbon filter, (bar), Q: flow rate, (m^3/hr.).

$$\Delta p_{carf} = 0.002154 \times Q^2 - 0.0005722 \times Q \quad (72)$$

where: $\Delta pcarf$: pressure drop across the cartridge filter, (bar), Q: flow rate, (m^3/hr.).

Multi-media filter, activated carbon filter, cartridge filter and filtration system experimental results, as shown in Figure 42.

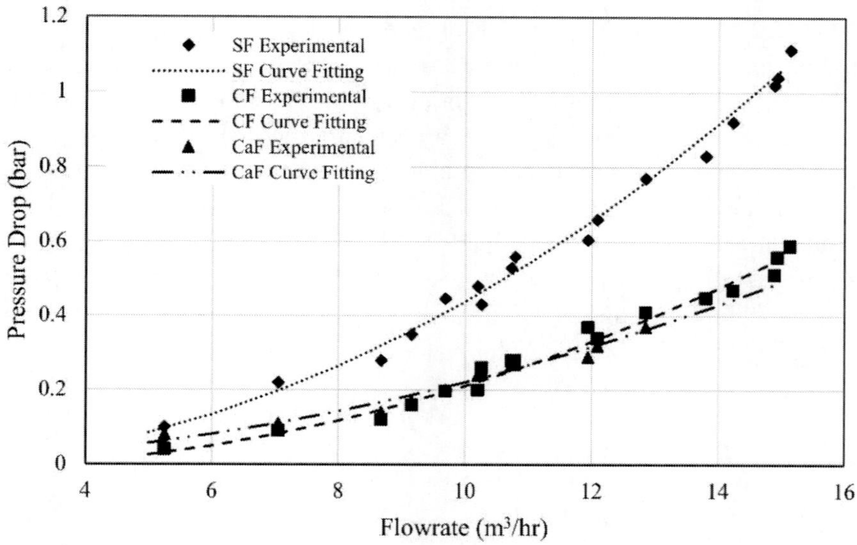

Figure 42. Multi-Media Filter, Activated Carbon Filter, and Cartridge Filter Experimental Results.

Conclusion

Mathematical modeling has become an essential technique in RO system design and optimization. Through the use of mathematical equations, accurate simulations can be achieved, enabling the prediction of the performance of RO systems under different operating conditions. This makes it possible for researchers and engineers to establish important performance indicators, including water quality and pressure losses. Detailed examination of particular factors affecting the performance of reverse osmosis units, coupled with developments in algorithmic methods and computer programming, greatly improves the potential for system performance optimization and greatly lowers operating costs.

Maintaining feed water quality is crucial to the effective operation of RO systems applied in desalination operations. The pre-treatment process, particularly through direct filtration using a variety of media methods, plays a significant role in reducing harmful material that could otherwise compromise the integrity of the RO membranes as well as the overall system efficiency. The layout of custom-designed testing hardware to rate individual filtration components guarantees the truthful measurement of their performance at various operating conditions. With the utilization of high-tech measurement instruments and a comprehensive data acquisition system, the experimental rig offers reproducible data demonstrating the interrelation between the pump characteristics and operating parameters. The agreement between the manufacturer's rating, experimental findings, and empirical correlations reflects the efficiency of the filtration system and its components. Additional focus on the optimization of pre-treatment processes will result in better life and performance of RO systems for water desalination processes.

Chapter 7

Reverse Osmosis Unit Modeling and Characteristic Measurement

Abstract

This chapter introduces a theoretical model for simulating and modeling a reverse osmosis (RO) unit from a hydrodynamic point of view. Overall mathematical models for all of the unit's component's, i.e., high-pressure pump, and membrane element, have been developed. Theoretical research was conducted to ascertain the pressure fluctuations associated with the piston pump, under conditions with and without the inclusion of a hydraulic accumulator, over a variety of flow rates and under different water salinity concentrations. Research was conducted to predict piston pump's dynamic characteristics and assess RO membrane performance under normal and abnormal operating conditions. In addition, experimental tests were conducted on an operational RO system, where pressure and flow rate were quantified. Pressure changes generated by the pump at varying flow rates, in addition to the permeate flow of RO membrane at dissimilar seawater salinities, were evaluated experimentally. Experimental data collected from these experiments, along with published experimental data on RO membranes, were compared with the corresponding theoretical findings from the mathematical model of the unit for verification of the simulation program. Validated simulation model was utilized as a basis for conducting a predictive study aimed at examining the dynamic response of the piston pump and the RO membrane performance during normal and abnormal operating conditions.

Keywords: mathematical modeling, hydraulic, piston pump, concentration polarization, high-pressure system, recovery, operational conditions

1. Introduction

Mathematical modeling and optimization play a central role in the RO system design process. The advances in computational power in recent years have enabled the creation of more powerful software packages for meshing, geometric modeling, and visualization of results, as well as the application of sophisticated analysis algorithms [193]. A mathematical model replicates the performance of the RO system for different operating conditions, and one can thereby predict performance parameters like output water quality and pressure drop. While mathematical equations cannot always completely represent the real situation, they are available as useful systems involving a set of simplifying assumptions [178, 179]. Typically, RO mathematical models are derived from the energy balance and mass balance principles.

Energy balance equations describe the pressure drop and energy consumption of the system, and the mass balance equations characterize the flow of water and solute molecules through the RO membrane. Mathematical modeling of RO systems enables both operating and design parameters to be optimized by predicting the effects of variations in conditions, for example, feed water temperature, pressure, and salinity. Furthermore, modeling may assist in optimizing membrane selection and configuration, predicting the effects of fouling on system performance, and estimating the operational lifespan of the RO system.

To desalination researchers and engineers, RO mathematical modeling is an emerging tool, providing deep insight into system behavior and optimization [179, 181, 194].

In this chapter, the overall mathematical model and corresponding measuring parameters of RO unit will be explained, detailing the equations governing each part. Simulation program will integrate the developed mathematical equations to build an overall simulation model of the RO unit. Parts of pre-treatment unit and RO unit are shown in Chapter 6, Figure 37.

2. Reverse Osmosis Unit Static Characteristics Mathematical Model

RO unit consists of high-pressure pump, feed header, RO membranes element, hydraulic accumulator, reject header, and throttling valve, taking into consideration that the used high-pressure pump and RO membrane are piston type, spiral wound type.

2.1. High-Pressure Pump (HPP)

The utilized HPP for RO desalination is described as a positive displacement machine of axial piston type. The reciprocating movement of the piston causes reciprocating fluctuations in flow rate and pressure. These fluctuations mainly occur due to mixing of fluid from several pistons being expelled into one outlet port of the pump. Figure 43 is a cross-sectional view of the cylinder of the pump consisting of a piston, stuffing box, inlet and outlet valves, and manifolds associated with them [195]. The flow rate to the high-pressure pump is equal to the outflow from the filters. The equation of the pump flow rate characteristic can be developed as presented earlier. The exit flow rate from the filters is the input of the high-pressure pump.

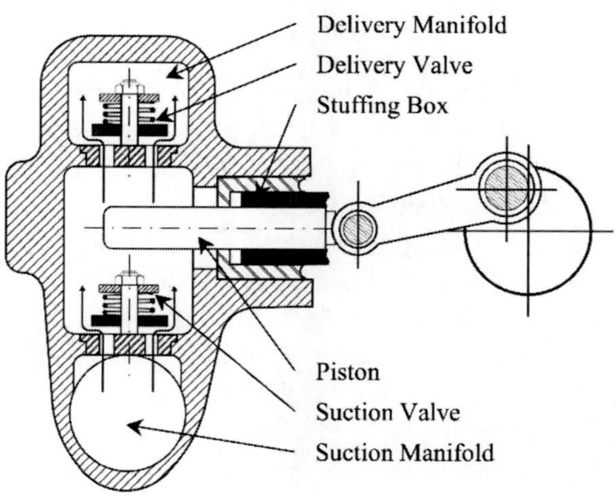

Figure 43. Cross-Section View of Fluid Cylinder.

The pump's real characteristics could be determined from the experimental measurements. The pump's real characteristic helps in the determination of the pump dynamics as the HPP used in RO desalination is a positive displacement type using an axial piston arrangement.

The reciprocating motion of the piston results in variations in flow rate and pressure. These flow fluctuations mainly occur due to the combination of fluid discharged by multiple pistons into a single outlet port of the pump. The input flow rate of the high-pressure pump is equal to the outflow rate from the filters. The governed equation which represents the flow rate of the pump may be expressed as shown in Equation (34) and (74).

$$Q_{pp} = V_g \times n \times \xi_m \times \xi_b \ \eta_{vol} \tag{73}$$

$$V_g = \frac{\pi}{4} D^2 \times s \times z_p \tag{74}$$

where: Q_{pp}: piston pump flowrate, (m^3/s); V_g: geometric volume, (m^3/rev); η_{vol}: volumetric efficiency; Z_p: number of Pistons; D: piston diameter, (m); n: pump revolutions per second, (rps); ξ_m: electric motor slip coefficient; ξ_b: belt slip coefficient; and s: piston stroke, (m), [182].

The working characteristics of a piston pump are based on a set of independent variables, the most important being volumetric losses. These losses result from several causes like leakage, inefficient filling, aeration, and cavitation. In pumps working with low-viscosity liquids, volumetric losses appear in the delivery system due to leaks between the fixed and moving members.

The actual delivery of the pump at a specified speed and fluid viscosity is obtained by subtracting the volumetric losses from the theoretical displacement. The main parameters that influence these volumetric losses are viscosity, μ; pressure, p; clearances, c; and speed, n. [182] the pump volumetric losses are a function of ($p.c^3/\mu$) apart from incomplete filling and cavitation effects. Since the pump operates with positive pressure in the suction line, it can be concluded that the pump is fully primed, and cavitation is not present. The volumetric losses can be represented as described [9].

The performance characteristics of piston pumps are a function of a group of independent parameters, among which volumetric losses hold paramount significance. Volumetric losses are affected by leakage, partially filled volume, aeration, and cavitation. In processes with low-viscosity fluids, volumetric losses in the discharge system of the pump are primarily caused by leakage between moving and stationary parts. The actual pump capacity at any speed and fluid viscosity is determined by dividing the theoretically displaced volume and volumetric losses due to leakage. many parameters influencing volumetric losses majorly involving pressure (p), clearances (c), viscosity (μ), and pump speed (n). Volumetric losses are also defined as a function of ($p \cdot c^3/\mu$), together with the influences of incomplete filling and cavitation. The pump has been specifically developed for operation in conditions of positive pressure in the suction line, which demands the full filling of the pump along with the elimination of cavitation. Therefore, volumetric losses can be quantitatively defined as described in Equation (75).

$$q = q_{SB} + q_V + q_L \tag{75}$$

where: q: volumetric losses, (m^3/s); q_{SB}: leakage through the stuffing box, (m^3/s); q_V: back flow across the valves, (m^3/s); and q_L: internal leakage, (m^3/s).

Volumetric losses could be defined as a percentage of the suction capacity. while the pump delivery flow rate at an operational point is a function of the suction flow rate and the operating pressure. Then the employed equation that may explain the piston pump performance could be expressed as shown in Equation (76).

$$Q_{HPP} = Q_{pp} - k \cdot p \tag{76}$$

where: Q_{HPP}: actual flowrate of the high-pressure pump, (m^3/s); k: volumetric losses coefficient, ($m^5/N.s$); Q_{PP}: piston pump flowrate, (m^3/s); and p: discharge pressure, (Pa).

2.2. Reverse Osmosis Membrane

RO process is well known as a pressure-driven separation process that has been used to extract water from saline solutions through a membrane [196]. In practice, two classes of membranes are considered for commercial applications: asymmetric and thin film composite membranes. Both classes have a porous substructure with a dense surface layer with high salt rejection while simultaneously allowing permeate passage [197]. The membrane used in RO is a molecular filter with pores of about 0.0006 microns, allowing only water molecules to pass through, and removing up to 99% of dissolved minerals. Although dissolved salts can diffuse, their solubility in the membrane is less than that of water; hence, separation occurs. The driving forces for this separation are the feed pressure and concentration differentials.

Many factors can majorly affect the performance of membranes including operated pressure, feed water characteristics such as temperature and concentration, and the determined recovery rate. For instance, higher implemented feed pressure increases permeate flux but decreases salt rejection, higher feed water temperature raises the permeate flow and salt rejection, higher feed concentration decreases permeate flux but raises salt rejection and higher recovery increases permeate flux and salt rejection. High-pressure pump pressurizes the feed water to the membrane system, with the outflow rates separated into two different streams: permeate and brine (concentrate) flow rate. The membrane element design is specifically made to meet the needed design parameters to optimize the water treatment process. In the current model, there are two sets of parallel membrane elements, each set consisting of two series elements. The two series elements are contained within a single pressure vessel as indicated in Figure 44.

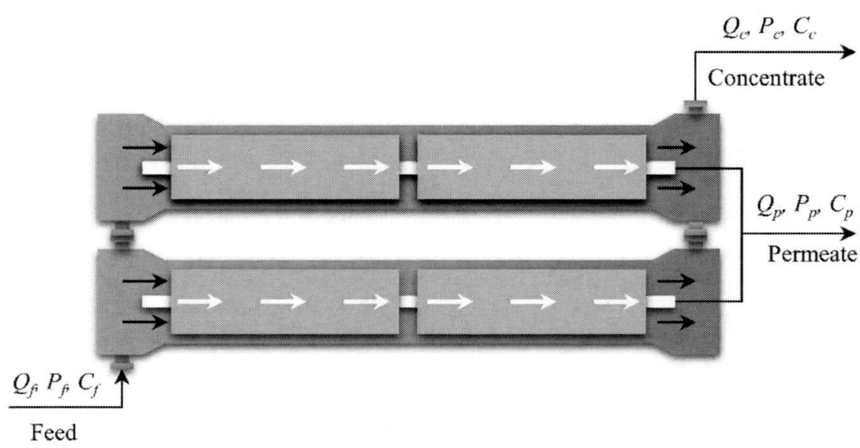

Figure 44. Membrane Arrangement.

The performance of RO element is defined by the flux of permeate at a given pressure, and by the salt rejection in the permeate water. In a mathematical way, the permeate flowQ_p through a RO element is directly proportional to the wetted surface area of the membrane (*S*), times proportionality constant (*A*) which is the membrane permeability coefficient, times the net permeation driving force$(\Delta p - \Delta\pi)$. Figure 45 displays the pressure distribution along the RO membrane and the way it impacts the permeation flow rate. The representing equation has the form as shown in Equation (77).

$$Q_p = A \times S \times (\Delta p - \Delta\pi) \tag{77}$$

where: Q_p: Permeate flowrate, (m^3/day), A : Membrane permeability coefficient, S : *Membrane surface area, (m^2),* Δp: Pressure drop in the membrane, (bar), and $\Delta\pi$: Osmotic pressure drop in the membrane, (bar).

The salt flux, N_A is proportional to the salt concentration difference between both sides of the membrane, as described in Equation (78). The proportionality constant is the salt diffusion coefficient or B_S value,

$$N_A = B_S \times (C_{fc} - C_p) \tag{78}$$

where: C_{fc} :Average concentration of feed-concentrate, (ppm) and C_p: Average permeates concentration, (ppm).

For examination of the performance of a membrane system in a general manner, average values of several parameters are considered, such as feed pressure and permeate quality as a function of feed salinity, temperature, permeate flow rate, and membrane elements' number. Where feed pressure is given instead of permeate flow rate, then the permeate flow rate may be determined.

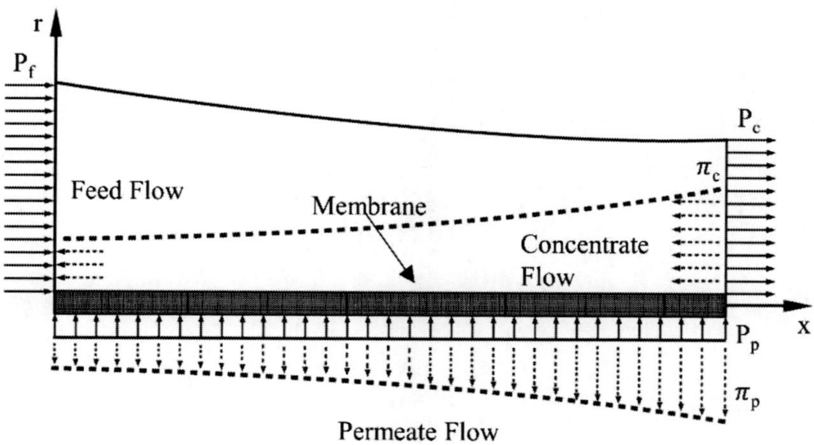

Figure 45. Pressure Distribution in a Membrane Element.

The master equation for membrane performance is the overall system permeate flow rate equation, as described in Equation (79) that has a series of correction factors, including the Temperature Correction Factor (*TCF*) for membrane permeability, the membrane fouling factor FF, and the number of elements in the systemN_E The permeate flow rate equation is then given:

$$Q_P = N_E \times S_E \times \bar{A}(\bar{\pi}) \times TCF \times FF \times \{\Delta p - \Delta\pi\} \tag{79}$$

where: Q_p: Permeate flowrate, GPD (gallon/day), A: Membrane permeability coefficient, S_E: Membrane surface area, (ft^2); $\bar{A}(\bar{\pi})$: Average membrane permeability, (G.ft^2.D/psi), Δp: Pressure drop in membrane, (psi), and $\Delta\pi$: Osmotic pressure drop in membrane, (psi).

The following are the procedures for calculating the driving pressure, average membrane permeability, temperature correction factor, and the fouling factor for a differential element of the system.

2.2.1. Driving Pressure

Driving pressure for a membrane element includes: Pressure drop in membrane (Δp) and Osmotic pressure drops in the membrane ($\Delta\pi$).

2.2.1.1. Pressure Drops in the Membrane, Δp.

Assuming that the pressure along the membrane element length is distributed linearly, Figure 46 shows the pressure drop distribution along the membrane on the feed-concentrate side. The pressure distribution acting along the membrane element is assumed to be the average value of feed and concentrated pressures, p_{fc}, as shown in part (a) of Figure 46. The pressure drops across the membrane thickness, Δp, is the difference between the average feed-concentrate pressure, p_{fc}, and the permeate pressure, p_p, as shown in part (b) of Figure 46. The pressure drops across the membrane, Δp, can be calculated using Equations (80), (81), (82), and (83) [182].

$$\Delta p = p_{fc} - p_p \tag{80}$$

$$p_{fc} = p_f - \frac{\Delta p_{fc}}{2} \tag{81}$$

$$\Delta p_{fc} = p_f - p_c \tag{82}$$

$$\Delta p = p_f - \frac{\Delta p_{fc}}{2} - p_p \tag{83}$$

where: p_{fc}: Average feed-concentrate pressure, (Pa), p_p: Permeate pressure, (Pa), p_f: Membrane feed pressure, (Pa), p_c: Membrane concentrate pressure, (Pa) and Δp_{fc}: Pressure drop between feed and concentrate, (Pa).

Pressure drop along the membrane length expresses the membrane performance; therefore, the pressure drop along the membrane length is considered in the simulation program. The pressure drops along the membrane length, Δp_{fc}, could be calculated using Equations (84), and (85).

$$\Delta p_{fc} = 0.01 \times n_e \times \bar{q}_{fc}^{1.7} \tag{84}$$

$$\bar{q}_{fc} = \frac{Q_f + Q_c}{2} \tag{85}$$

where*:* n_e: Number of elements in series, $\bar{q}_{fc}$: Arithmetic average of feed-concentrate flowrate, *(GPM),* Q_f: System feed flowrate, (GPM) and Q_c*: System concentrate flowrate, (GPM).*

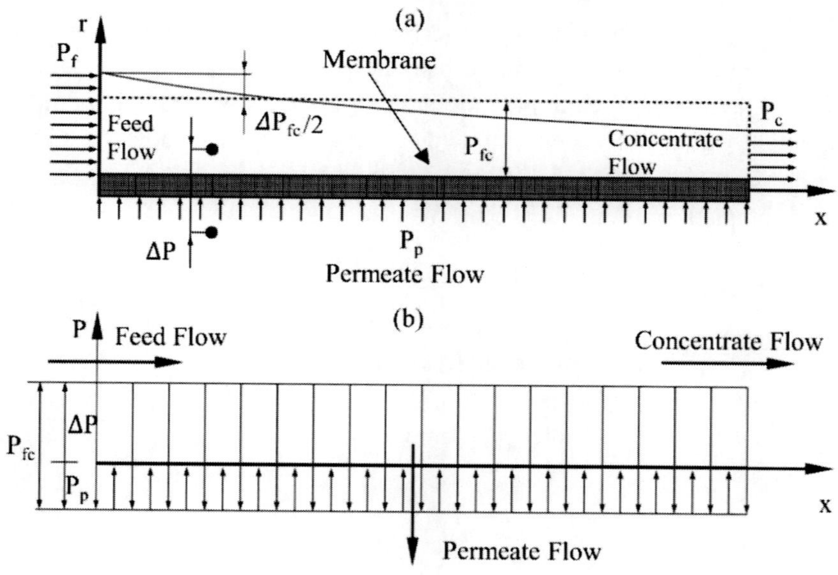

Figure 46. (a) Pressure Distribution Along the Membrane. (b) Pressure Drop Across the Membrane.

2.2.1.2. Osmotic Pressure Drops in Membrane

Assuming the osmotic pressure along the membrane length is distributed linearly. Figure 47 shows the osmotic pressure drop distribution along the membrane length on the feed-concentrate side. The osmotic pressure distribution acting along the membrane element is assumed to be the average value of feed and concentrated osmotic pressures, π_{fc}. The osmotic pressure on the membrane side is increased due to the contamination of the salt ions at the membrane boundaries. The layer containing the ions of salts is defined as the *Concentration Polarization layer*. This layer affects the osmotic pressure value by a factor called average concentration polarization factor, *pf*, as shown in part (a) 0f Figure 47.

Concentration polarization (CP) is a phenomenon that occurs due to the accumulation of excess salts or dissolved molecules on the membrane surface, consequently affecting the driving force, and leading to a level off of the actual flux data. CP is inherent to all the membrane filtration processes and the resistance to solvent flow increases to decrease permeate flux. Permeate flux reduction is induced by solute accumulation and osmotic pressure increase on the membrane surface.

Convective and diffusive solute transport in opposite directions causes a concentration profile normal to the membrane. For the present model, the osmotic pressure distribution acting across the membrane element is assumed to be the average feed-concentrate osmotic pressure. The osmotic pressure difference across the membrane, $\Delta\pi$, is the osmotic pressure of the permeate, π_p, and the average feed-concentrate osmotic pressure,$\bar{\pi}$, as indicated by part (b) of Figure 47. The osmotic pressure differences across the membrane, $\Delta\pi$, can be obtained from the Equation (86).

$$\Delta\pi = \bar{\pi} - \pi_p \tag{86}$$

where: $\bar{\pi}$: Average concentrate-side osmotic pressure, (bar) and π_p: Permeate-side osmotic pressure, (bar).

The average concentrate-side osmotic pressure, $\bar{\pi}$, could be calculated using Equation (87).

$$\bar{\pi} = \pi_f \times \frac{C_{fc}}{C_f} \times pf \tag{87}$$

where: π_f: Feed water osmotic pressure, (bar), C_f: Feed water concentration, (ppm) *and pf*: Average concentration polarization factor.

The feed water osmotic pressure, Average concentration of feed concentrate, and Average concentration polarization factor could be calculated as follows:

Feed Water Osmotic Pressure: The osmotic pressure of feed water depends on the molar concentration of dissolved ions in the feed water, and the feed water temperature. The feed water osmotic pressure, π_f, could be calculated as introduced as shown in Equation (88).

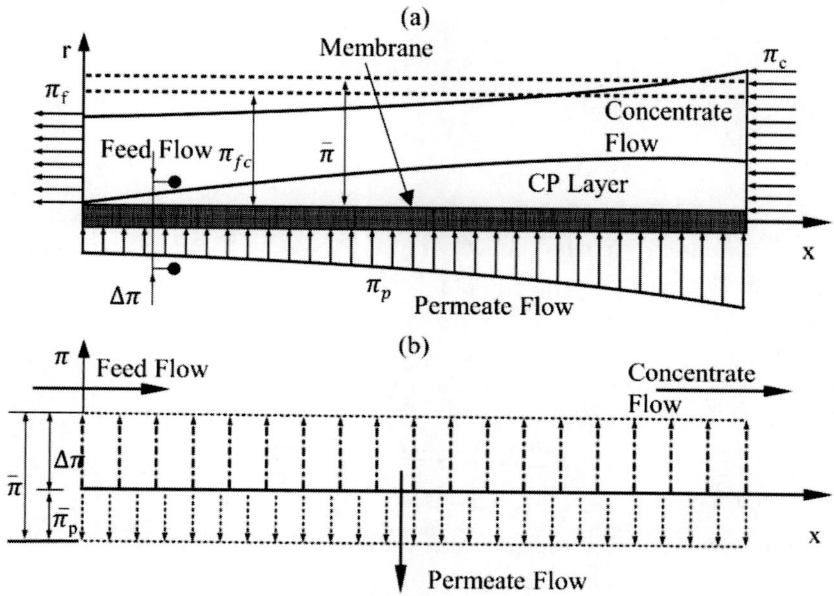

Figure 47. (a) Osmotic Pressure Distribution Along the Membrane Length. (b) Osmotic Pressure Drops Across the Membrane Thickness.

$$\pi_f = 1.12 \times [273 + T] \times \sum m_j \tag{88}$$

where: T: Feed water temperature, (°C) and $\sum m_j$: Summation of all molar concentrations of ionic species.

Average concentration of feed-concentrate: Assuming the water concentration along the membrane length is distributed linearly. Figure 48 shows the concentration distribution along the membrane on the concentrate side. The average concentration of feed-concentrate, C_{fc}, could be calculated using Equation (89).

$$C_{fc} = \frac{C_f + C_c}{2} \tag{89}$$

where concentration distribution along the membrane element is assumed to be the average value of feed and concentrated water concentration, C_{fc}.

Average concentrations polarization: With real substances, a layer is deposited at the membrane surface affects the driving force. The actual flux data shows a leveling off as a concentration layer builds up at the membrane surface. Average concentration polarization factor can be calculated using Equation (90), and (91).

$$pf = e^{[0.7\times\bar{Y}_i]} \tag{90}$$

$$\bar{Y}_i = 1 - [1 - Y]^{1/ne} \tag{91}$$

where: $\bar{Y}_i$: Average element recovery ratio, *Y*: System recovery ratio and n_e: Number of elements in series.

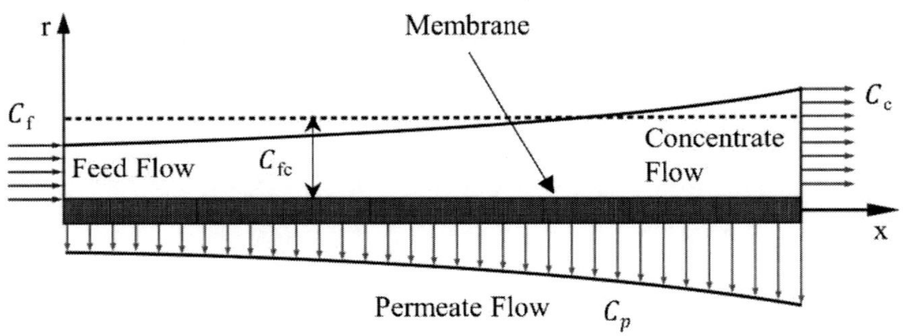

Figure 48. Average Concentration of Feed-Concentrate.

The average permeate-side osmotic pressure, π_p, could be calculated using Equation (92).

$$\pi_p = \pi_f \times (1 - \bar{R}) \tag{92}$$

where: $\bar{R}$: Average fractional salt rejection for the system.

2.2.2. Average Membrane Permeability

The Average membrane permeability, $\bar{A}(\bar{\pi})$, can be calculated as introduced in using Equation (93).

$$\begin{aligned}
\bar{A}(\bar{\pi}) &= [0.125]/0.069; && \bar{\pi} \le 25\ Psi \\
\bar{A}(\bar{\pi}) &= [0.125 - 0.011 \times [\bar{\pi} - 25]/35]/0.069; && 25 \le \bar{\pi} \le 200\ Psi \\
\bar{A}(\bar{\pi}) &= [0.070 - 0.0001 \times [\bar{\pi} - 200]]/0.069; && 200 \le \bar{\pi} \le 400\ Psi
\end{aligned} \tag{93}$$

2.2.3. Temperature Correction Factor

Feed water temperature is among the environmental conditions that have an important effect on the operation of RO systems. Temperature, in particular, affects the porosity of the membrane and hence permeate flow rate; as temperature increases, permeate flow increases, while as temperature decreases, flow decreases. In order to calculate permeate flow rate at different temperatures correctly, *TCF* must be included in the permeate calculation. *TCF* may be calculated as described in Equation (94).

$$\begin{aligned} TCF &= e^{[2640\times\{1/298-1/(273+T)\}]}; \quad T \geq 25°C \\ &= e^{[3480\times\{1/298-1/(273+T)\}]}; \quad T \leq 25°C \end{aligned} \tag{94}$$

2.2.4. Fouling Factor

The RO membrane elements' surface is susceptible to fouling because feed water contains a wide range of foreign materials that can include various contaminants such as metal oxide hydrates, calcium precipitates, organic matter, and biological contaminants. Fouling in this context refers to the deposition of any layers on the membrane surface such as scaling effects. To reduce the contamination of membrane surfaces, pre-treatment of feed water before the RO process is essential. This reduction is attained by adopting proper pretreatment systems and optimizing operating parameters such as permeate flow rate, pressure, and recovery ratio of permeate water. Indeed, Surface fouling of the membrane may be caused by several factors, including incorrect design of the pretreatment system, pretreatment process upsets, poor material selection (for piping, pumps, etc.), failure of dosing chemicals equipment, incorrect flushing practice after shutdowns, absence of effective operational control, continuous precipitate accumulation over extended time periods (most notably barium and silica), changes in feed water composition, or biological fouling of the feed water.

The "fouling factor" is used in calculations to account for the impact of fouling on the membrane surface. The factor varies from 0.1 to 1, depending on feed water quality and pretreatment process quality. The employed system of the desalination plant in the present chapter is *spiral wound* membrane, *Thin-Film Composite*, and its model is *SW30HR-380, 8"* Seawater RO Elements. The previously presented equations and factors are used to simulate the membrane performance using a simulation program. The membrane model has been obtained by using the iteration method. The developed membrane model has been obtained using the empirical relations, for another membrane manufacturer the empirical relations for $\bar{A}(\bar{\pi})$, Δp_{fc} and *TCF* should be used.

3. Reverse Osmosis Unit Dynamic Characteristics Mathematical Model

A dynamic model has been developed to evaluate the dynamic performance of the high-pressure pump component of the RO desalination plant. The dynamic model has been obtained for both the high-pressure pump without membrane, HPP, and the high-pressure pump loaded with membrane (HPPM).

HPP includes the pump, the accumulator, and the throttling valve. HPPM includes the pump, the accumulator, the membrane, and the throttling valve. The dynamic models have been obtained by applying the characteristic equations of the system components. The equations that represent each part will be shown separately. The simulation program reveals the link between the equations to form the simulation model of the HPP and HPPM systems.

4. High-Pressure Pump Without Membrane Mathematical Model

High-pressure pump without membrane consists of three main components: the high-pressure pump, the accumulator, and the throttling valve. Figure 49 presents HPP components.

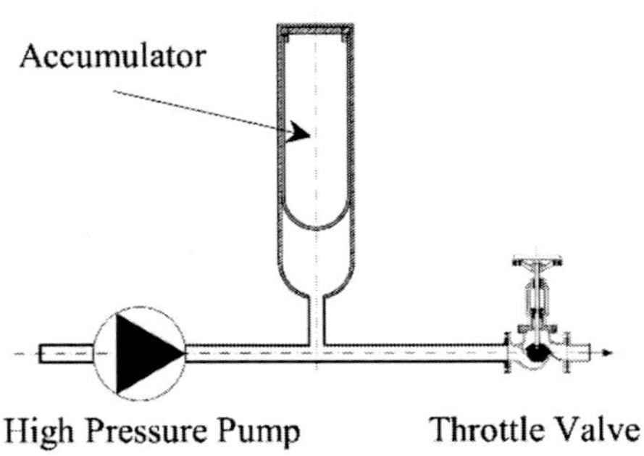

Figure 49. High-pressure Pump Without Membrane.

A description of the characteristic equation of each component will be presented as follows:

4.1. High-Pressure Pump Characteristic Equations

In order to study the dynamic performance of the high-pressure pump, the equations governing the fluid motions inside the piston chamber should be derived. The piston chamber is composed of four parts, the piston, the suction valve, the delivery valve, and the chamber, as shown in Figure 43.

4.1.1. Piston

The piston displaces a given volume of fluid in each stroke. The piston transmits the force that develops the pressure. The delivered flow of fluid throughout the high-pressure pump (piston pump) occurs in a transitory dynamic cycle called a *pumping cycle.* The source of this cycle is the reciprocating linear movement of the piston.

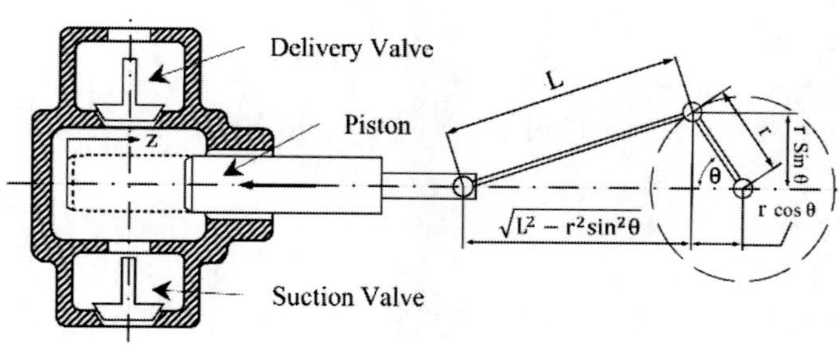

Figure 50. Slider Crank Mechanism.

In Figure 50, *r* is the radius of the crank in meters, *L* is the length of the connecting rod in meters, *C* equals *L/r*, and ω equals *(2π/60)* × rpm. Thus, the linear movement of the piston is related to the above-mentioned parameters is presented in [182] as described in Equation (95).

$$Z = r[1 - \cos\theta] + L\left(1 - \sqrt{1 - \frac{r^2}{L^2}\ (\sin\theta)^2}\right) \tag{95}$$

where: $\bar{R}$: Average fractional salt rejection for system.

Consider the process of suction stroke in a single acting pump, neglecting r w.r.t. L i.e., $r/L \approx 0$. Thus, on its way from the extreme left position rightwards, the piston covers the distance Z, as presented in Equations (96), ,(97) and (98).

$$Z = r - r\cos\theta \tag{96}$$

$$\theta = \omega\ t \tag{97}$$

$$Z = r - r\cos\omega\ t \tag{98}$$

where: ω: Angular velocity, (rad/s), and t: Time (sec).

Piston velocity of the determines the discharged flowrate from the cylinder. The approximate velocity of the piston can be calculated using Equation (99).

$$v_p = \frac{dZ}{dt} = r\omega\ \sin\omega\ t \tag{99}$$

where: v_p: Piston velocity, (m/s).

Piston velocity times piston area gives the volume rate of sucked fluid by the piston, which can be calculated using Equation (100).

$$Q_{pis} = A_p \times r\omega\ \sin\omega\ t \tag{100}$$

where: Q_{pis}: Piston flowrate, (m^3/s), and A_p: Piston cross section area (m^2).

The pump flowrate for five pistons results from the superposition of the volume rate of liquid drawn by each piston with ($2\pi/5$) angle shift for each piston, which can be calculated using Equation (101) or (102).

$$Q_{pis_j} = A_p\ r\omega\ \ \sin(\omega t + \frac{(j-1)}{z_p}2\pi) \tag{101}$$

where: $Q_{pis\,j}$: Piston flowrate, (m^3/s) and j: 1-5.

The total pump flowrate Q_{hpp}

$$Q_{hpp} = \sum_{j=1}^{5} Q_{pis_j} \tag{102}$$

4.1.2. Suction Valve

The suction valve is a one-way-valve of motion depending on the stroke of the piston. The suction valve is of disk (plate) type. The disc is made of Titanium. The valve consists of the seat and the plate. The plate movement is controlled by a spring and retainer (limiter); Figure 51 shows the check valve assembly.

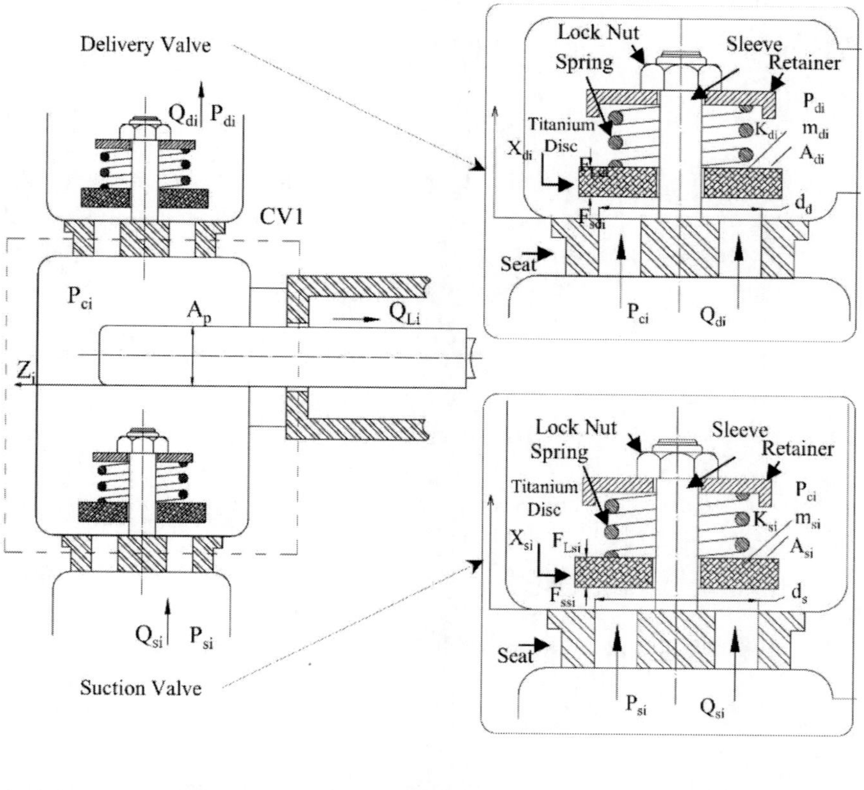

Figure 51. Piston Chamber Control Volume, Suction Valve and Delivery Valve Assembly.

Consider the suction valve parameters as in Figure 51, the equations that describe the suction valve could be written as follows:

A. Flowrate through the suction valve: The flowrate through the suction valve is throttled through variable area $A(x_{Si})$. The flowrate through suction valve could be written as described in Equation (103) and (104).

$$Q_{Si} = C_{dS} A(x_{Si}) \sqrt{\frac{2(P_{Si} - P_{Ci})}{\rho}} \tag{103}$$

$$A(x_{Si}) = x_{Si} \times \pi d_s \tag{104}$$

where: Q_{Si}: Flowrate through suction valve (m^3/s), C_{dS}: Suction valve discharge coefficient, $A(x_{Si})$: Suction valve throttling area (m^2), P_{Si}: Suction pressure (Pa), P_{Ci}: Chamber pressure (Pa) and ρ: Water density (kg/m^3).

The pump suction flowrate through the suction valve equals to the superposition of flowrates through the individual suction valves. The suction flowrate is given by Equation (105).

$$Q_S = \sum_{i=1}^{5} Q_{Si} \tag{105}$$

B. Equation of motion of moving parts: The suction valve moving parts are moving under the action of pressure forces, spring force, inertia force, seat reaction force, and limiter reaction force. The displacement of the suction valve springs is not the same for all suction valves at the same time. The suction valve spring displacement is related to the difference between pressure pulsations at each chamber and the pressure at the suction line of the pump. Neglecting the gravity force w.r.t. the pressure force, the equation of motion of the suction valve could be written as described in Equation (106).

$$m_{Si}\ddot{x}_{Si} + f_{Si}\dot{x}_{Si} + k_{Si}(x_{Sio} + x_{Si}) = (P_{Si} - P_{Ci})A_{Si} + F_{SSi} - F_{LSi} \tag{106}$$

where: m_{Si}: Suction valve disc mass (kg), f_{Si}: Suction valve moving parts friction coefficient (N.s/m), k_{Si}: Suction valve spring stiffness (N/m), x_{Sio}: Pre-compression length of suction valve spring (m), x_{Si}: Displacement of suction valve spring (m), A_{Si}: Effective area of suction valve disc (m^2), F_{SSi}: Seat reaction force of suction valve (N), F_{LSi}: Limiter reaction force of suction valve (N) and $_i$: Suction valve number (1 to 5).

C. Seat and limiter reaction forces: The dynamic components are urged to move up and withdraw towards the valve seat by virtue of pressure changes. This motion is constrained at the valve seat by the physical limitations of the valve body material, which results in the generation of a reactive force against the motion at the seat. The equation for the seat reaction force can be expressed as described in Equation (107).

$$F_{SSi} = \begin{cases} 0 & \rightleftarrows x_{Si} > 0 \\ k_{S1}x_{Si} + f_{S1}\dot{x}_{Si} & x_{Si} \leq 0 \end{cases} \tag{107}$$

where: k_{S1}: Stiffness of moving parts at seat of suction valve (N/m) and f_{S1}: Suction valve seat damping coefficient (N.s/m).

At the other direction of motion, the valve is limited mechanically by a limiter. The limiter reaction force could be written as described in Equation (108).

$$F_{LSi} = \begin{cases} k_{S2}(x_{Si} - x_{Sm}) + f_{S2}\dot{x}_{Si} & (x_{Si} - x_{Sm}) \geq 0 \\ 0 & (x_{Si} - x_{Sm}) < 0 \end{cases} \quad (108)$$

where: x_{Sm}: Maximum displacement of suction valve spring (m), k_{S2}: Stiffness of moving parts at limiter of suction valve (N/m) and f_{S2}: Suction valve limiter damping coefficient (N.s/m).

4.1.3. Delivery Valve

The delivery valve has the same specifications as the suction valve. Considering the delivery valve parameters as in Figure 51, the equations that describe the delivery valve could be written as follow:

A. Flowrate through the suction valve: The flowrate through the delivery valve is throttled through variable area $A(x_{di})$. The flowrate through suction valve could be written as described in Equation (109), and (110).

$$Q_{di} = C_{dd}A(x_{di})\sqrt{\frac{2(P_{Ci}-P_{di})}{\rho}} \quad (109)$$

$$A(x_{di}) = x_{di} \times \pi d_d \quad (110)$$

where: Q_{di}: Flowrate through delivery valve (m^3/s), C_{dd}: Delivery valve discharge coefficient, $A(x_{di})$: Delivery valve throttling area (m^2) and P_{di}: Delivery pressure (Pa).

The pump delivery flowrate is given by Equation (111).

$$Q_d = \sum_{i=1}^{5} Q_{di} \quad (111)$$

B. Equation of motion of moving parts: The delivery valve moving parts are moving under the action of pressure forces, spring force, inertia force, seat reaction force, and limiter reaction force. The displacement of the delivery valve springs is not the same for all delivery valves at the same time. The delivery valves spring displacement is related to the difference between pressure pulsations at each chamber and the pressure at the discharge line of the pump. Neglecting the gravity force w.r.t. the pressure force, the equation of motion of the suction valve could be written as described in Equation (112).

$$m_{di}\ddot{x}_{di} + f_{di}\dot{x}_{di} + k_{di}(x_{dio} + x_{di}) = (P_{Ci} - P_{di})A_{di} + F_{Sdi} - F_{Ldi} \tag{112}$$

where: m_{di}: Delivery valve disc mass (kg), f_{di}: Delivery valve moving parts friction coefficient (N.s/m), k_{di}: Delivery Valve spring stiffness (N/m), x_{dio}: Initial compression distance of delivery valve spring (m), x_{di}: Displacement of delivery valve spring (m), A_{di}: Effective area of delivery valve disc (m^2), F_{Ldi}: Limiter reaction force of delivery valve (N) and F_{Sdi} : Seat reaction force of delivery valve (N).

C. Seat and limiter reaction forces: The moving parts move up and back to the valve seat following the change in pressure fluctuations. At the valve seat, the direction of the motion is limited mechanically by the valve body material and results in development of a counter seat reaction force. The seat reaction force could be written as described in Equation (113).

$$F_{Sdi} = \begin{cases} 0 & x_{di} > 0 \\ k_{d1}x_{di} + f_{d1}\dot{x}_{di} & x_{di} \leq 0 \end{cases} \tag{113}$$

where: k_{d1}: Stiffness of moving parts at seat of delivery valve (N/m) and f_{d1}: Delivery valve seat damping coefficient (N.s/m).

The limiter reaction force could be written as described in Equation (114).

$$F_{Ldi} = \begin{cases} k_{d2}(x_{di} - x_{dm}) + f_{d2}\dot{x}_{di} & (x_{di} - x_{dm}) \geq 0 \\ 0 & (x_{di} - x_{dm}) < 0 \end{cases} \tag{114}$$

where: x_{dm}: Maximum displacement of delivery valve spring (m), k_{d2}: Stiffness of moving parts at limiter of delivery valve (N/m) and f_{d2}: Delivery valve limiter damping coefficient (N.s/m).

4.1.4. Piston Chamber

Consider the piston chamber parameters as in Figure 51, the equations that describe the piston chamber could be written as follows:

A. Continuity equation applied to piston chamber: The continuity equation applied to the control volume CV1 in the piston chamber as shown in Figure 51. could be written as described in Equation (115).

$$A_p\dot{z}_i + Q_{Si} - Q_{di} - Q_{Li} = \frac{(V - A_p z_i)}{B}\frac{dP_{ci}}{dt} \tag{115}$$

where: Q_{Li}: Piston leakage flowrate (m^3/s), V: Chamfer initial volume (m^3), B: Bulks modules of water (kN/m^2) and z_i : Displacement of piston (m).

B. Leakage flowrate: The leakage flowrate (Q_{Li}) around the piston is linearly proportional to the pressure difference across the piston head as described in Equation (116).

$$Q_{Li} = \frac{P_{ci} - P_{atm}}{R_i} \tag{116}$$

where: R_i: Resistance to leakage, ($N.s/m^5$).

4.2. Accumulator Characteristic Equation

A diaphragm accumulator operates based on gas compressibility, i.e., nitrogen, to accumulate hydraulic energy. this diaphragm acts as a separator that maintains the separation between the gas and seawater inside the accumulator. When there is a pressure rise in the hydraulic circuit, water is pushed into the accumulator, thereby compressing the gaseous phase. Conversely, when circuit pressure decreases, the gas is expanded, and this provides an opportunity for water to be released back into the hydraulic system.

A. Continuity equation applied to the accumulator: The continuity equation applied to the control volume CV2 as shown in Figure 52, could be written as described in Equation (117).

$$Q_d - Q_v = Q_a \tag{117}$$

where: Q_v: Valve flowrate (m^3/s) and Q_a: Accumulator flowrate (m^3/s).

B. Ideal gas law: The diaphragm of the accumulator may be examined based on the principles of the ideal gas laws. As shown in Figure 52, the working stages are classified into four categories: (a) pre-charged state, (b) minimum working pressure, (c) working pressure, and (d) maximum working pressure. Taking an isothermal process, the state equation relevant to the accumulator, based on the ideal gas law, can be written as described in Equation (118).

$$P_o \times V_o^{\alpha} = P_1 \times V_1^{\alpha} = P_d \times V_a^{\alpha} = P_2 \times V_2^{\alpha} \tag{118}$$

where: P_o: Gas pre-charge pressure, (Pa), V_o: Effective gas volume of the accumulator, (m^3), P_2: Maximum working pressure, (Pa), V_2: Gas volume at P_2, (m^3), P_1: Minimum working pressure, (Pa), V_1: Gas volume at P_1, (m^3), V_a: Gas volume in accumulator at P_d, (m^3) and α: Polytropic exponent.

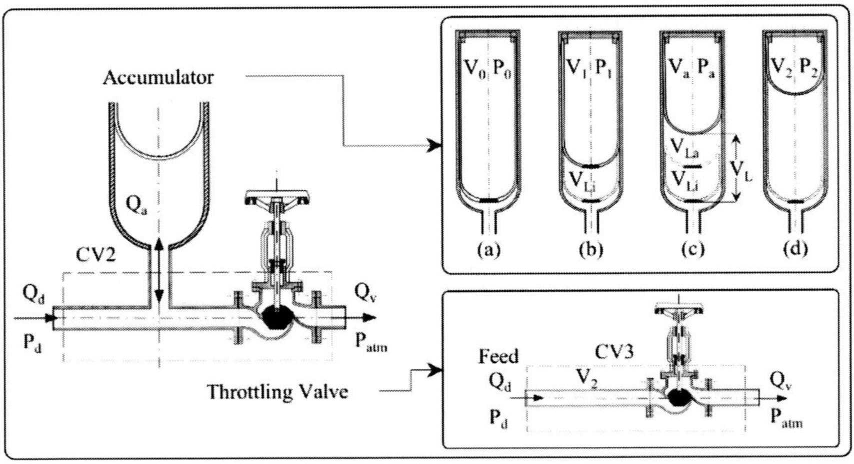

Figure 52. Control Volume Around Accumulator, Throttle Valve and Accumulator Working Phases.

C. Accumulator flowrate: The accumulator flowrate (Q_a) could be written as described in Equation (119), (120), (121), and (122).

$$Q_a = \frac{dV_{La}}{dt} \tag{119}$$

$$V_L = V_{Li} + V_{La} \tag{120}$$

$$V_{Li} = V_o - V_1 \tag{121}$$

$$V_a = V_o - V_L \tag{122}$$

where: V_{La}: Volume of the liquid entering the accumulator at P_d, (m^3), V_{Li}: Initial volume of the liquid at the accumulator at P_1, (m^3) and V_L: Volume of the liquid inside the accumulator at P_d, (m^3).

4.3. Throttle Valve Characteristic Equation

Throttling valve is used to control the pump delivery pressure. The throttling valve used type is a globe valve.

A. Flowrate through throttle valve: The discharge flowrate from the pump enters the throttle valve. The throttle valve is opened to the atmospheric pressure at its exit. The throttle valve flowrate could be written as described in Equation (123).

$$Q_v = \frac{A_v}{\sqrt{\xi_v}}\sqrt{\frac{2(p_d - p_{atm})}{\rho}} \tag{123}$$

where: Q_v: Throttle valve flowrate (m^3/s), ξ_v: Throttle valve hydraulic losses coefficient, A_v: Throttle valve throttling area (m^2) and p_{atm}: Atmospheric pressure (Pa).

B. Continuity equation applied to the throttle valve: The continuity equation applied to the control volume CV3 as shown in Figure 52, could be written as described in Equation (124).

$$Q_d - Q_v = \frac{V_T}{B}\frac{dP_d}{dt} \tag{124}$$

where: V_T: Water volume from pump output to throttle valve (m^3).

The dynamic characteristic of the high-pressure pumping unit is described mathematically by the non-linear differential equations derived for all components. A simulation program based on these equations has been developed by using computer program.

5. High-Pressure Pump Loaded by Membrane Mathematical Model

The high-pressure pump loaded by membrane consists of six components: the high-pressure pump, the accumulator, the feeding header, the membrane, the rejection header, and the throttle valve as shown in Figure 53. The high-pressure pump and the accumulator equations are as presented earlier. Description of the remaining components equations could be written as follows:

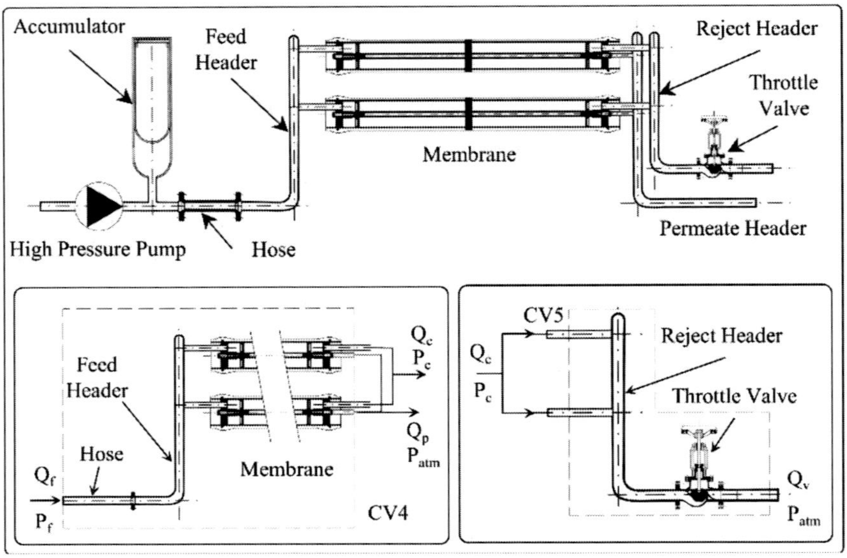

Figure 53. High-Pressure Pump Loaded by Membrane Components, Feed Header and Membrane Control Volume, and Reject header and throttling valve control volume.

5.1. Feed Header and Membrane Characteristic Equation

Feed header is used to direct feed water from the high-pressure pump to the membrane system. The feed water inside the membrane is separated into two different channels: the first channel is reserved for permeate water, while the other channel is for concentrated water. The operations of both the feed header and the membrane are governed by some mathematical relations which can be shown using the following equations:

5.2. Feed Header and Membrane Applied Continuity Equation

The continuity equation applied to the control volume CV4 as shown in Figure 53, could be written as described in Equation (125), and (126).

$$Q_f - Q_c - Q_p = \left(\frac{V_h}{B_e} + \frac{V_{fh}}{B}\right)\frac{dP_f}{dt} \tag{125}$$

$$B_e = \frac{1}{1/B + 5D_h/4Eh} \qquad (126)$$

where: V_h: Water volume inside feed water hose (m^3), V_{fh}: Water volume at feed water header (m^3), D_h: Hose diameter (m), E: Bulks modules of hose material (kN/m^2), h: Hose wall thickness (m), and B_e: Equivalent bulks modules of water and hose material (kN/m^2).

The pressure loss in feed header may be considered as local loss type. The losses in pressure could be neglected w.r.t. the operating pressure of the system.

5.3. Reverse Osmosis Membrane

The equations that describe the flowrate in membrane could be written as follows:

A. Concentrate flowrate: The concentrate flowrate out of the membrane could be calculated from Equation (84), and (85) after rearranging to have the form as described in Equation (127).

$$Q_C = 2 \times \left(\frac{P_f - P_C}{0.01 \times n_e}\right)^{1/1.7} - Q_f \qquad (127)$$

B. Permeate flowrate: The permeate flowrate out of the membrane has been described by Equation (79).

5.4. Reject Header and Throttling Valve

The rejection header is employed for gathering the reject water from the high-pressure side of the membrane's output, leading it through the throttle valve. The throttle valve is governed for controlling the feed pressure entering the membrane. The dynamics of the rejection header and the throttling valve can be represented by the following equations:

A. Continuity equation applied to reject header and throttling valve: The continuity equation for the reject header and throttling valve applied to the control volume CV5 as shown in Figure 53. could be written as described in Equation (128).

$$Q_c - Q_v = \frac{V_R}{B}\frac{dP_c}{dt} \quad (128)$$

where: V_R: Water volume between membrane output and throttle control valve (m^3).

B. Flowrate through throttling valve: The delivery flowrate exit from the membrane enters into the throttle valve. The throttle valve flowrate could be written as described in Equation (129).

$$Q_v = \frac{A_v}{\sqrt{\xi_v}}\sqrt{\frac{2(P_C - P_{atm})}{\rho}} \quad (129)$$

The pressure loss in reject header may be considered a local loss which could be neglected relative to the operating pressure of the system.

A simulation program has been developed based on the above-mentioned equations by using computer program to simulate the characteristics of the high-pressure pump loaded by the membrane. The obtained model for the RO unit components should be validated by experimental measurements on the real RO plant.

6. Static Characteristic Measurement

A specific test rig has been employed for the two components of the RO unit; these rigs are strategically designed to place measuring devices at optimal positions for evaluating the static characteristics of both components under several operating conditions. The test rig encompasses two parts: the operating hydraulic system, which contains the selected component under evaluation, and a set of measuring devices for data gathering. Extensive measurements were performed for the two RO components the high-pressure pump and the RO element as described below.

6.1. High-Pressure Pump Test Rig

The hydraulic circuit utilized in testing the high-pressure pump. Equipment involved in testing the performance characteristics of the high-pressure pump consists of a storage tank, a booster pump, two flow meter transducers (one positioned upstream and the other downstream of the pump), two high-pressure transducers (similarly positioned), a hydraulic accumulator, a globe valve, and a set of pipes and hoses. The measurements are taken by noting the suction pressure, delivery pressure, suction flow rate, and delivery flow rate over a series of pump pressures, namely from 0 to 70 bar. The data obtained enable the plotting of the pump delivery pressure-flow rate curve. This pressure-flow rate characteristic is needed for the determination of pump leakage at any operating pressure. Experimental measurements reveal that the volumetric losses at an operating pressure of 70 bar are around 4.65%. The pump characteristic equation, obtained by curve fitting of experimental data, is presented below in Equation (130).

$$Q_{HPP} = 10.384 - 0.0069 \times p \quad (130)$$

where: Q_{HPP}: actual flowrate of the high-pressure pump, (m^3/hr.), and p: discharge pressure, (bar).

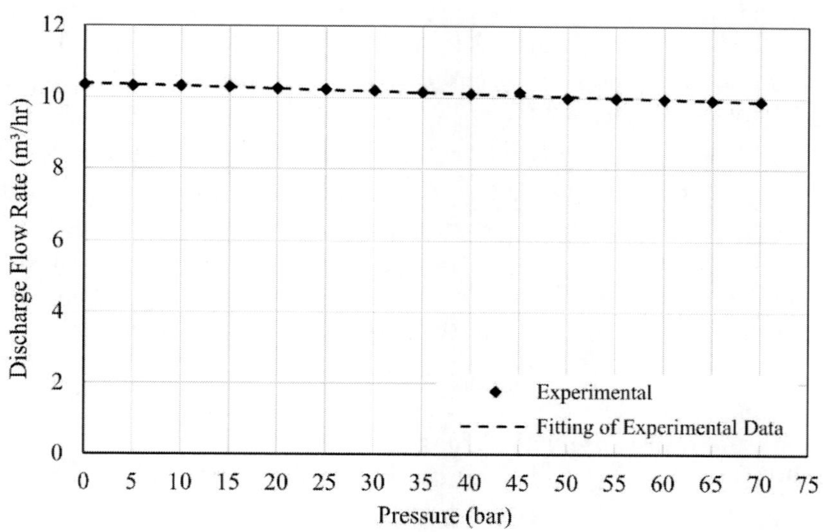

Figure 54. Experimental Results of the High-pressure Pump.

Figure 54 shows the fitting curve and the experimental results of the high-pressure pump. These results will be used later in the determination of pump dynamic model.

6.2. Membrane Test Rig

The hydraulic circuit utilized in testing the high-pressure pump. Equipment involved in testing the performance characteristics of the high-pressure pump consists of a storage tank, a booster pump, two flow meter transducers (one positioned upstream and the other downstream of the pump), two high-pressure transducers (similarly positioned), a hydraulic accumulator, a globe valve, and a set of pipes and hoses. The hydraulic circuit employed in the testing of the membrane. Two flow meter transducers, two high-pressure transducers, and one conductivity transducer are the measuring instruments utilized in ascertaining the membrane's properties. It passes through a closed system, starting from the storage tank holding an adequate volume of artificial seawater (NaCl equivalent), then through the booster pump, the filtration unit, and the high-pressure pump before it reaches the membrane inlet section. The permeate and concentrate water lines go back to the storage tank. The high-pressure transducers are connected to the feed (inlet) and concentrated sides of the membrane. The conductivity transducer and flow meter transducer are connected to the permeate side of the membrane. A flow meter transducer is connected to the suction line of the piston pump (booster pump outlet) for measuring the feed water flow rate. The measurement was conducted to quantify the total dissolved solids and temperature of the synthetic seawater. The feed pressure was adjusted to a specific value utilizing the throttle valve. Concentrate pressure, feed flow rate, permeate flow rate, and permeate conductivity were recorded at feed pressures between 30 and 60 bar. The measurements were carried out three times with three varied concentrations of artificial seawater, i.e., 21,000 ppm, 28,900 ppm, and 33,000 ppm of NaCl.

7. Dynamic Characteristic Measurement

7.1. Test Rig and Instrumentation

The same test setup used for static characteristic measurement is used for the measurement of the dynamic characteristic of the piston pump. For this purpose, the data acquisition system is set to a higher sampling rate to be able to acquire the high-frequency pressure ripples of the piston at different operating pressures and conditions. The measurement sensor used is a calibrated high-pressure transducer. Data acquisition system consists of a data acquisition card, a terminal board, and a personal computer. Analog-to-digital (A/D) and digital-to-analog (D/A) converting facilities are offered by the data acquisition card. In particular, it has one digital-to-analog channel, programmable for either unipolar or bipolar operation, and eight analog-to-digital channels, each programmable for unipolar or bipolar operation and providing a 12-bit resolution per channel.

The card is installed within the personal computer and linked to the terminal board through a data cable. The A/D-D/A terminal board offers connection points for the analog and digital input/output signals with two digital-to-analog terminals and sixteen analog-to-digital terminals that support single-ended analog-to-digital connection. The terminal board is connected to the A/D-D/A card through a 25-pin flat cable. The A/D-D/A card is firmly mounted inside the motherboard of the personal computer. The computer runs the A/D-D/A card to collect data using special software, which also helps in collecting and saving data.

7.2. High-Pressure Pump Unit

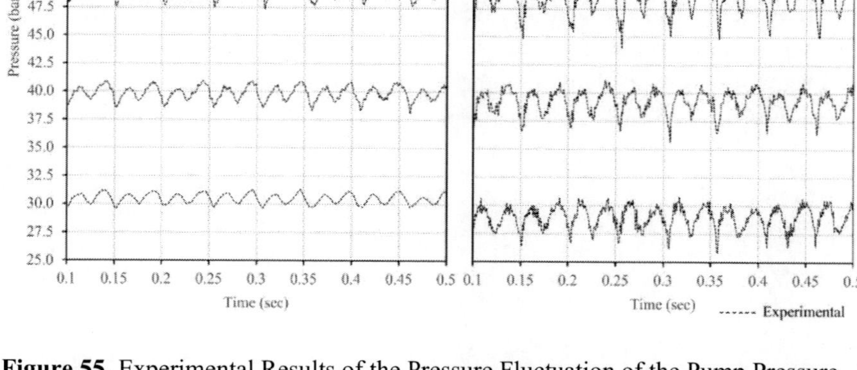

Figure 55. Experimental Results of the Pressure Fluctuation of the Pump Pressure Fluctuations, (a) With an Empty Accumulator, (b) Without Accumulator, at Working Pressures of 60.3, 50.8, 40, and 29.8 bar.

The hydraulic circuit used in the testing of the dynamic characteristics of the piston pump is analogous to those used in the testing of the static characteristics. The testing is done by recording the pressure fluctuations of the pump delivery line with and without the accumulator in the hydraulic circuit over the pressure range from 30 to 60 bar. The results acquired have been used to demonstrate the pressure variations in the pump. The experimental results are plotted in Figure 55. The experimental results of the piston pump with the empty accumulator included in the circuit are shown in Figure 55 part (a), whereas the results achieved from the piston pump when the accumulator is taken out of the circuit are shown in Figure 55 part (b). The experimental results show that the presence of the accumulator greatly reduces pressure fluctuations in the hydraulic circuit even when the accumulator is empty.

Conclusion

Mathematical modeling is considered an emerging tool in the design and optimization of RO unit. The use of mathematical equations gives accurate simulations that enable the prediction of the behavior of the RO unit under different conditions of operation; this enables the researcher or engineer to extract key performance metrics, which may include water quality and pressure losses. Detailed examination of certain factors affecting, RO unit, combined with new algorithms and software developments, increases our ability to optimize system efficiency and reduce operating costs substantially. The characterization of RO unit critical components high-pressure pump, feed header, RO membranes, hydraulic accumulator; reject header, and throttling valve, shows how each element contributes to the general effectiveness of the RO unit performance and sustainability.

Chapter 8

Reverse Osmosis Unit Simulation Model Validation

Abstract

This chapter presents an experimental study on a reverse osmosis (RO) system. The investigation comprised extensive flow rate and pressure behavior measurements. More specifically, pump pressure variations were measured at various flow rates, along with the permeate pressure of the RO membrane at varying seawater salinity levels. To validate the hypothesis of the simulation program used, a comparison analysis was made on experimentally supplied data and literature data of the RO membrane, with theoretical estimates using the theory model of the unit. The validated simulation model also enabled the prediction of high-pressure pump performance and RO membrane performance during normal and non-normal plant operating modes. RO plant simulation model was employed to devise an attainable operation diagram of the intended desalination plant, considering feed water temperature and salinity fluctuations.

Keywords: mathematical modeling, hydraulic, fluid, piston pump, membrane element, concentration polarization, high-pressure system, reverse osmosis recovery ratio, reverse osmosis operational conditions

1. Introduction

Sophisticated numerical simulations enable quantitative descriptions of the flow field and concentration profiles of dissolved species in the unit cells of spacer-filled channels with high accuracy for the RO process. Nevertheless, a huge requirement to validate these simulation results exists to confirm their credibility in simulating the models of desalination. The process of validation plays a key role in developing strong models, as it offers crucial experimental evidence that underpins mechanism evaluation, model simplification, and parameter estimation.

In essence, this strategy is geared towards improving preliminary assessments of research and guidance and streamlining the use of time and financial resources [198-200]. This chapter is dedicated to the validation of simulation models of the RO unit and its components. It will discuss at length the mathematical equations that model each component separately. By combining these equations, the simulation program tries to develop an overall model of the RO unit as a whole.

2. High-Pressure Pump Validation

Theoretical analysis of pressure fluctuation in the piston pump was accomplished using our simulation model. The comparison of the experimental and theoretical findings at different pump operating pressures reveals that the frequency and amplitude of the pressure signals are in general agreement at all operating pressures. Interestingly, the experimental pressure pumping contains variation, which can be attributed to fluid flow dynamics in the system, e.g., the effect of pipe fixation, fittings, connections, and other mechanical components [201].

A comparison study of theoretical hypothesis and real system results under various testing conditions is provided for the high-pressure pump working in two cases, with an empty accumulator, and without accumulator are shown in Figure 56. This figure shows the simulation and experimental results for pump operating pressure ranges from 30 to 60 bar. It is obvious that these results validate the simulation program of the dynamic operation mode.

3. Membrane Simulation Model Validation

The simulation modeling for the membrane was verified through comparative analysis with the experiment. Figure 57 present comparisons of permeate flow rates and salt rejection performance from the theoretical hypothesis with experimental measurements at three levels of salinity (21,000, 28,900, and 33,000 ppm). The experiments were conducted on two-by-two configurations of Film Tec SW30HR-380 elements, which are designed specifically for spiral wound membranes for seawater application, at feed pressures of between 30 and 65 bar.

4. High-Pressure Pump Simulation

The high-pressure pump dynamic performance is characterized by the value of flowrate irregularity coefficient, ε_Q, as described by Equation (74) or pressure irregularity coefficient, ε_P, as described by Equation (132).

$$\varepsilon_Q = \frac{Q_{max} - Q_{min}}{Q_{av}} \times 100\% \tag{131}$$

where: ε_Q: Flowrate irregularity coefficient, (%), Q_{max}: Pump maximum flowrate, (m^3/s), Q_{min}: Pump minimum flowrate, (m^3/s) and Q_{av}: Pump average flowrate, (m^3/s).

$$\varepsilon_P = \frac{QP_{max} - P_{min}}{P_{av}} \times 100\% \tag{132}$$

where: ε_P: *Pressure irregularity coefficient, (%),* P_{max}: *Pump maximum pressure, (bar),* P_{min}: *Pump minimum pressure, (bar) and* P_{av}: *Pump average pressure, (bar).*

The results obtained from the mathematical representations of the high-pressure pump that simulate different failure modes concerning individual components of the pump are presented in Figure 58 and Figure 59. The results discussed pertain to a pump with an operating pressure of 60 bar and include the following scenarios: a high-pressure pump with a pre-charged accumulator; a high-pressure pump with a charged accumulator but experiencing a sticking delivery valve; a high-pressure pump with an empty accumulator and a sticking delivery valve; a high-pressure pump with a charged accumulator when there is considerable leakage in one chamber; a high-pressure pump with an empty accumulator in the presence of considerable leakage in one chamber; a high-pressure pump without an accumulator but with a sticking delivery valve; and lastly, a high-pressure pump without an accumulator but experiencing considerable leakage in one chamber. For the high-pressure pump coupled to a charged accumulator, the accumulator's charged pressure is 30 bar with a volume of 5 liters. For the condition where heavy leakage occurs in a particular chamber, a reduction in the resistance corresponding to the leakage phenomenon is witnessed.

Also, in order to address the issue of the sticking valve, the variable area *A(xdi)* utilized in Equation (103) described in Chapter 7 is changed to match a constant area valve to be determined in the revised Equation (133).

$$Q_{di} = C_{dd} A_{ST} \sqrt{\frac{2(P_{Ci} - P_{di})}{\rho}} \tag{133}$$

where: A_{ST}: Delivery valve sticking area (m^2).

To predict the off-design conditions, some cases of piston pump flowrate and pressure fluctuations have been simulated such as; charged accumulator at 30 bar, as shown in Figure 58 part (a), charged accumulator at 30 bar while one delivery valve is sticking, as shown in Figure 58 part (b), an empty accumulator while one delivery valve is sticking, as shown in Figure 58 part (c), charged accumulator at 30 bar while excessive leakage at one chamber occurs, are shown in Figure 58 part (d), empty accumulator while excessive leakage at one chamber, is shown in Figure 59 part (a), without accumulator while one delivery valve is sticking, is shown in Figure 59 part (b), and without accumulator while excessive leakage at one chamber, are shown in Figure 59 part (c).

4.1. High-Pressure Pump Simulation Results

The dynamic simulation model established in this research accurately replicates the variation of flow rate and pressure for a high-pressure pump when operating under fluctuating conditions. The results suggest that the operation efficiency of the pump can be described using the irregularity factor for flow rate ε_Q and that for pressure ε_P. Importantly, these irregularity factors are affected by defects in the components of the pump.

Table 5. Flowrate irregularity coefficients values at different failure conditions

Failure	Flowrate Irregularity Coefficient, ε_Q, (%)		
	Pump with accumulator		Pump Without Accumulator
	Charged	Empty	
No failure	15.45	14.7	13.6
One delivery valve is sticking	34.52	34.56	26.3
Excessive leakage at one chamber occurs	21.4	19.9	48.6

Table 5 presents the flow rate irregularity coefficients for various failure conditions, and Table 6 states the pressure irregularity coefficients for the same conditions. The experimental values of pressure irregularity were determined as 5.8% for the case when the accumulator was not charged and 15% in the absence of an accumulator. The simulated values of pressure irregularity match the experimental values directly, which implies that the simulation model can be accepted as a trustworthy tool for further theoretical research.

Table 6. Pressure irregularity coefficients values at different failure conditions

Failure	Pressure Irregularity Coefficient, ε_P, (%)		
	Pump with accumulator		Pump Without Accumulator
	Charged	Empty	
No failure	0.1	5.83	15.3
One delivery valve is sticking	1.845	21.99	47.1
Excessive leakage at one chamber occurs	0.7	11.9	20.2

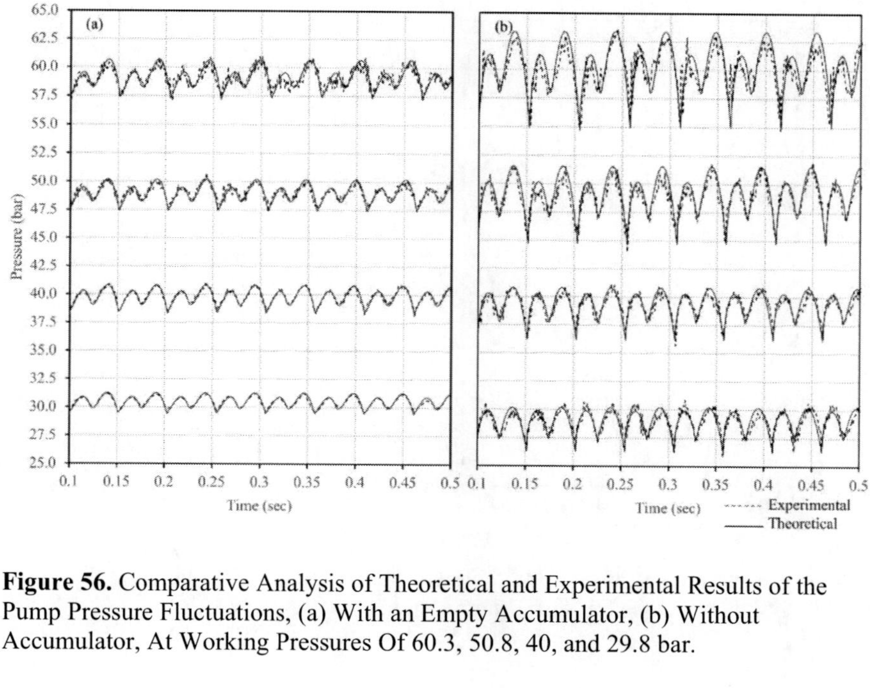

Figure 56. Comparative Analysis of Theoretical and Experimental Results of the Pump Pressure Fluctuations, (a) With an Empty Accumulator, (b) Without Accumulator, At Working Pressures Of 60.3, 50.8, 40, and 29.8 bar.

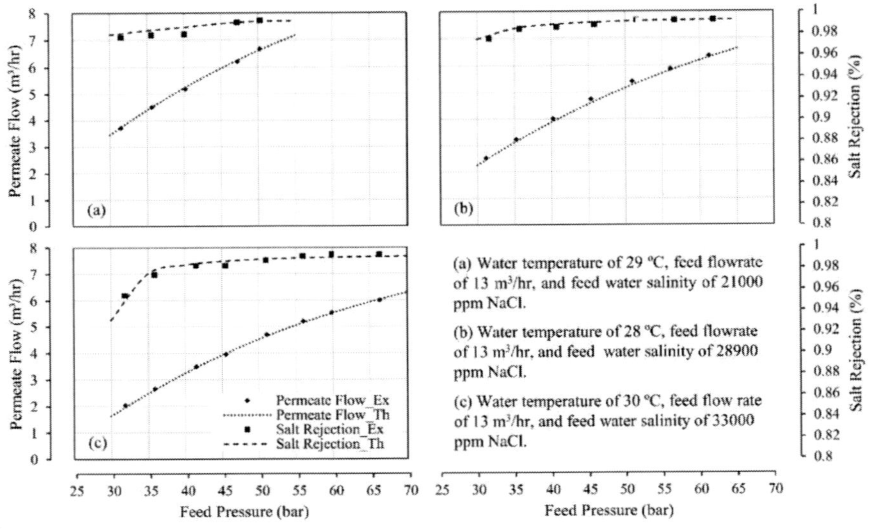

Figure 57. Comparative Analysis of Theoretical and Experimental Membrane Characteristics (Feed Pressure vs Permeate Flow) and (Feed Pressure vs Salt Rejection).

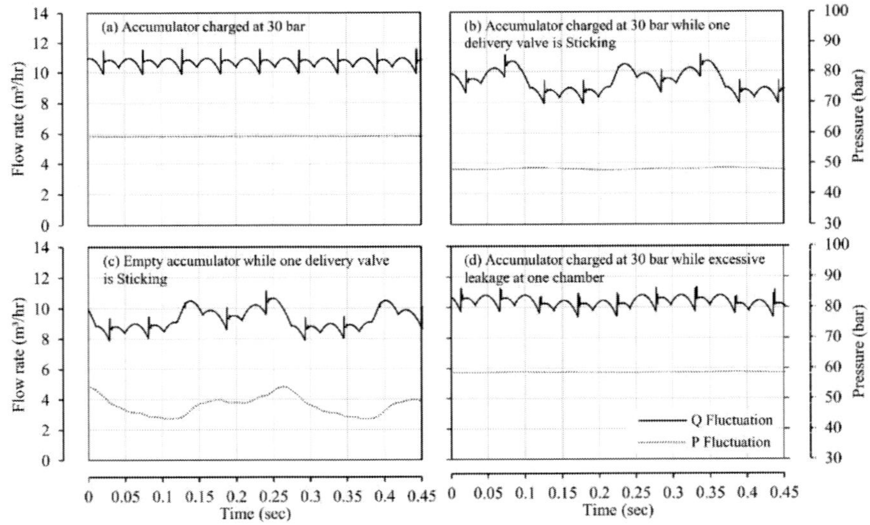

Figure 58. Simulation of Piston Pump Flowrate and Pressure Fluctuations at Different Failure Cases for Some Pump Components, (N = 227.4 rpm).

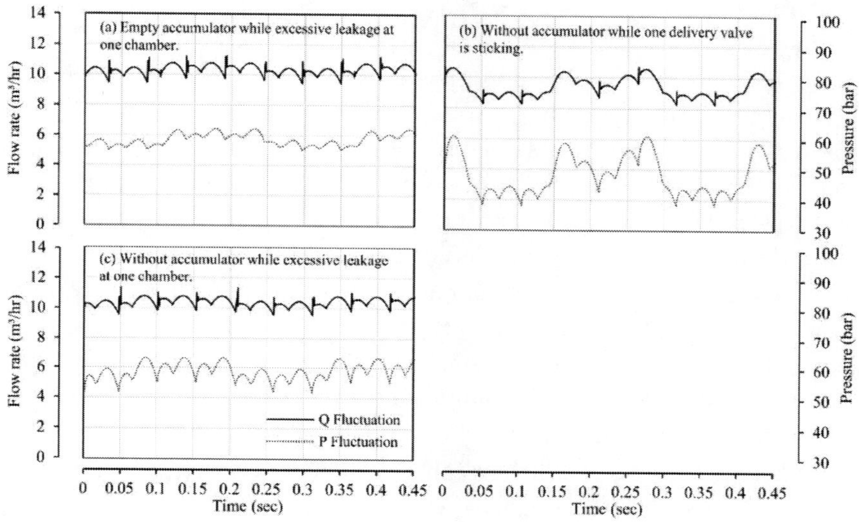

Figure 59. Simulation of Piston Pump Flowrate and Pressure Fluctuations at Different Failure Cases for Some Pump Components, (n = 227.4 rpm).

5. RO Membrane Simulation

The mathematical equations describing membrane behavior were solved using a simulation program. The model was established using advanced simulation software. To study the nature of the membrane, two models were created. The first model discusses the influence of feed pressure on the different operating parameters of the membrane. However, the second one discusses the impact of the projection recovery ratio on these operational parameters. Figure 60 to Figure 62 shows examples of the correlated variables obtained by theoretical simulations with the model. There are numerous considerations that affect the operational parameters impacting membrane performance in different ways. The simulation indicates the effect of variations in feed water salinity, actual temperature, and recovery ratio, as well as the exposed pressure on membrane.

Additionally, it describes the interaction of these conditions with the pressure drop along the exposed surface area of the membrane element and the flow rate of the permeate, both of which are important parameters that determine membrane efficiency. The variation in operating conditions significantly affects actual feed pressure, as well as the pressure drop along the direction of membrane length, and permeate flow rate.

5.1. Effect of Variation in Feed Water Salinity

Taking into consideration the adjustment of the Recovery ratio and feedwater temperature at constant values, thereby making it possible to analyze the impact of varying feedwater salinity on feed pressure at varying feed flow rates. The simulation indicates that, at some feed flow rates, feed pressure is directly proportional to feedwater salinity.

Feedwater salinity difference influences two of the key parameters that decide feed pressure: average membrane permeability ($\bar{A}(\bar{\pi})$) and feedwater osmotic pressure (πf). Since the salinity of feedwater is greater, the feedwater osmotic pressure (πf) is greater since the concentration of ions is high. Meanwhile, increasing feedwater salinity reduces average membrane permeability ($\bar{A}(\bar{\pi})$).

This impact results in a greater concentration of dissolved ions for every volume of feedwater, which makes feedwater more viscous. Consequently, there is a decrease in the membrane's average permeability, and to compensate for this decrease in permeability and achieve the required permeate flow through the membrane, there is a need to increase the pressure. Figure 60 part (a) indicates the influence of different feedwater salinity levels on the pressure exerted on the membrane, while Figure 60 part (b) depicts the interrelation of feedwater salinity changes, feed osmotic pressure, and average membrane permeability.

5.2. Influence of Variation of Feed Water Temperature

This chapter investigates the impact of varying feed temperatures on feed pressure at varying feed water flow rates, with constant feed water salinity and recovery ratio. Mathematical modeling results reveal that feed pressure and feed water temperature have negative attribution at constant feed flow rates. As shown in Figure 61 part (a), the feed pressure is lower at a higher feed water temperature in the membrane system. The reason for this kind of phenomenon may be related to the increase in the temperature correction factor (*TCF*) and feed water osmotic pressure (πf) when the temperature increases, along with the decrease in average membrane permeability ($\bar{A}(\bar{\pi})$). Contrary to expectations, the rise in osmotic pressure does not result in a proportional rise in feed pressure. Rather, the primary contribution to the decline in feed pressure comes from the rise in the temperature correction factor, and the decline in the average membrane permeability has an extremely minor contribution to this decline, as indicated in Figure 61 part (b).

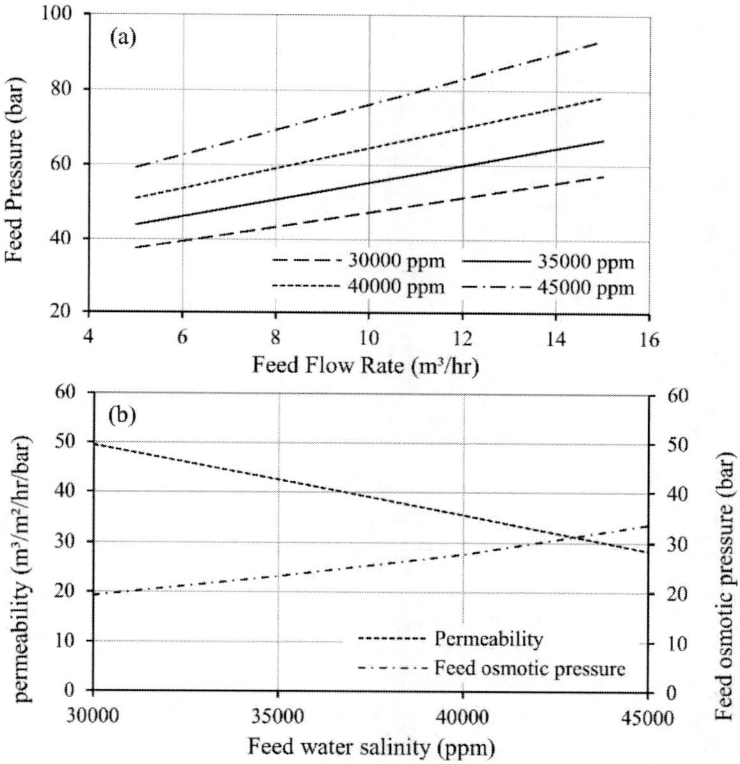

Figure 60. (a) Membrane Characteristic Curve (Feed Pressure vs. Feed Flowrate) Salinity Effect on Feed Pressure at Water Temperature of 22°C, and Recovery Ratio of 25%. (b) Influence of Feed Water Salinity on Permeability and Feed Osmotic Pressure at Constant Feed Flowrate.

5.3. Influence of Variation on Recovery Ratio

In order to analyze the influence of changing the recovery ratio on the feed pressure with different feed water flow rates, the values of salinity and temperature were maintained constant. Simulation outputs indicate that there is a linear proportionality between feed pressure and recovery ratio at a particular feed flow rate. Indeed, the operating pressure is found to rise with a rise in the recovery ratio.

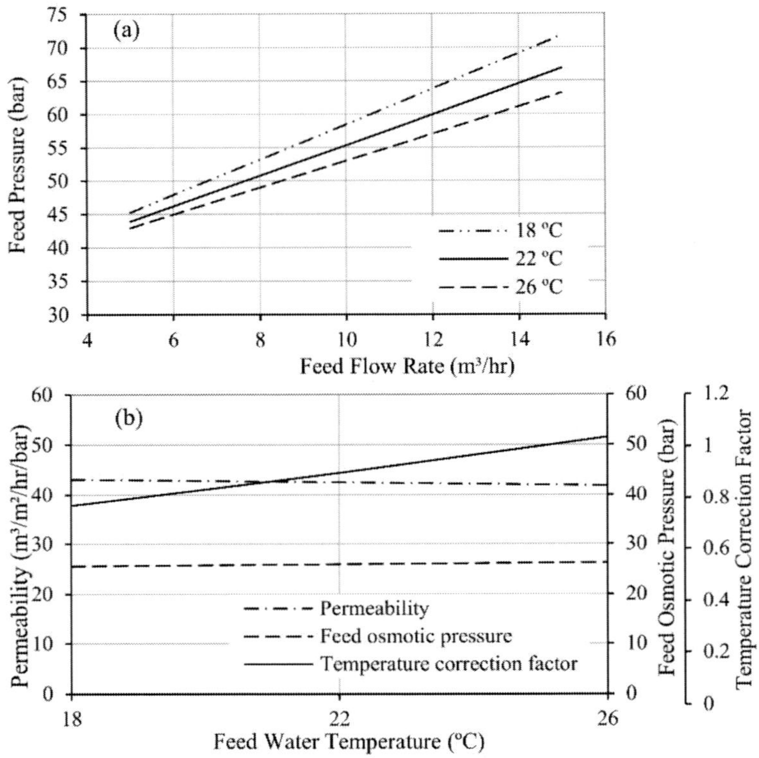

Figure 61. (a) Membrane Characteristic Curve (Feed Pressure vs. Feed Flowrate) Temperature Effect on Operating Pressure at Water Salinity of 35,000 ppm and Recovery Ratio of 25%. (b) Influence of Feed Water Temperature on Permeability, Feed Osmotic Pressure, and Temperature Correction Factor at Constant Feed Flowrate.

This result can be attributed to the reduction in the average concentrate-side system pressure drop (ΔP_{fc}), which results from a reduction in the average feed-concentrate flow rate (Q_{fc}) and a corresponding reduction in average membrane permeability ($\bar{A}(\bar{\pi})$). When the average feed-concentrate flow rate (Q_{fc}) is reduced, the ion concentration in the membrane boundary layer rises. Therefore, the average concentration polarization factor (*pf*) also rises and with it, the average concentrate-side concentration (C_{fc}) rises. This change leads to the reduction in membrane permeability$\bar{A}(\bar{\pi})$). Figure 62 reflect the relation between different recovery ratios (i.e., the ratio of permeate flow rate to feed flow rate) as shown in part (a) of the figure and the effects they exert respectively on feed pressure, mean membrane permeability, and pressure drop as shown in part (b) of the figure.

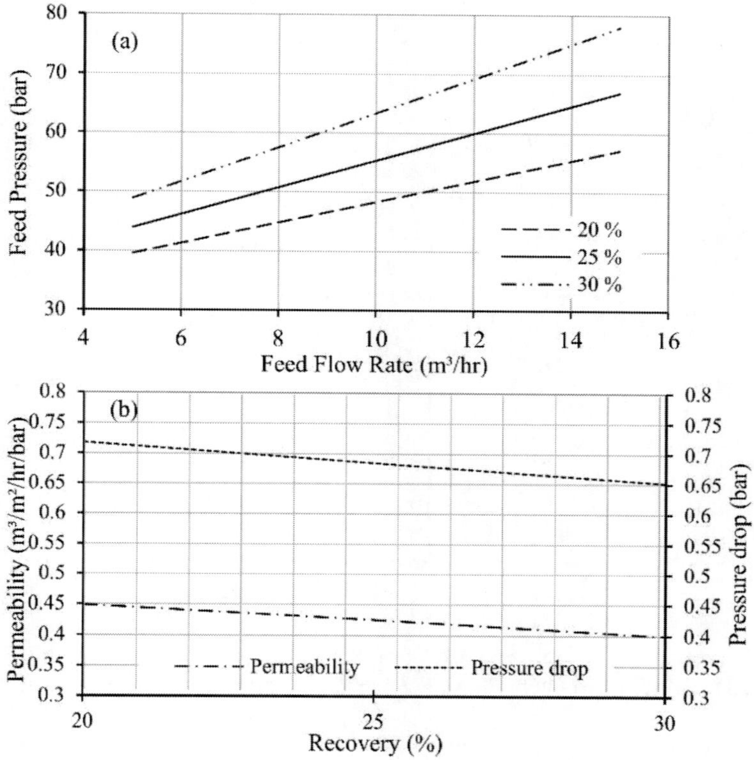

Figure 62. (a) Membrane Characteristic Curve (Feed Pressure vs. Flowrate) Recovery Ratio Effect on Operating Pressure at Water Salinity of 35,000 ppm, and Water Temperature of 22ºC. (b) Influence of Recovery Ratio at Permeability and Pressure Drop at Constant Feed Flowrate.

5.4. Membrane Performance

Permeate flux and salt rejection are key parameters for the quantitative as well as qualitative estimation of membrane performance. Permeate flux is described as the rate of passage of permeate through a unit surface area of membrane, classically quantitated in terms of liters per square meter per hour ($l/m^2 \cdot h$). Salt rejection is the proportion of the solute concentration that a membrane successfully impedes from being present in feed water. There are several factors that have a significant influence on membrane performance. These are feed water salinity, system recovery ratio, feed water pressure, and feed water temperature. The effect of these parameters on permeate flux and salt rejection is illustrated in Figure 63.

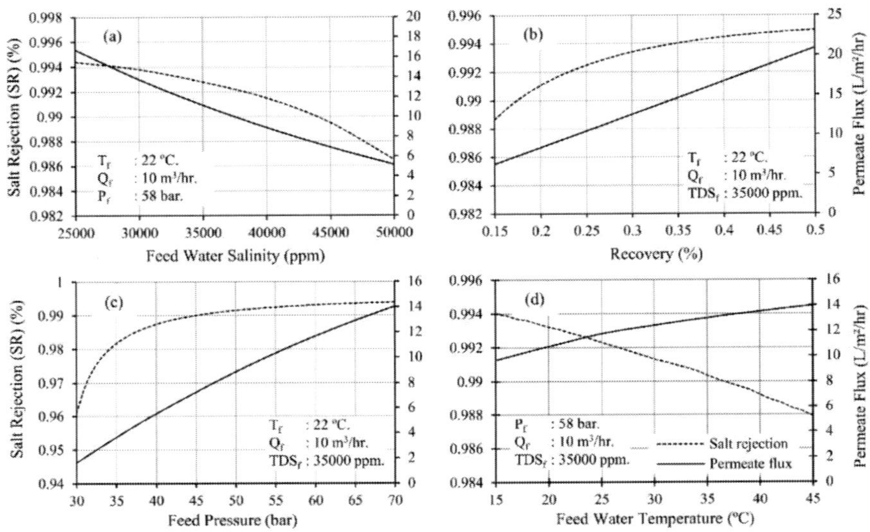

Figure 63. Membrane Characteristic Curve (Salt Rejection and Permeate Flux vs. Feed Water Salinity, Recovery Ratio, Feed Pressure and Feed Water Temperature).

5.4.1. Effect of Feed Water Salinity on RO Membrane Performance

The simulation results confirm that the rise in feedwater salinity leads to both a decreasing trend in permeate flux and salt rejection. The decrease in permeate flux has a linear correlation with the rising values of feedwater salinity. However, the decrease in salt rejection is a nonlinear phenomenon. This decrease in salt rejection is accompanied by a rise in membrane permeability. Despite the enhanced permeability of the membrane, the permeate flux actually declines due to increased osmotic pressure, leading to a reduction in net driving pressure. Figure 63 part (a) shows the effect of feedwater salinity on permeate flux and salt rejection.

5.4.2. Influence of System Recovery Ratio on RO Membrane Performance

The results of the simulation show that the system recovery ratio enhancement is positively correlated with both permeate flux and salt rejection. Specifically, the permeate flux against the system recovery ratio shows roughly a linear trend. Nevertheless, the enhancement of salt rejection versus the system recovery ratio demonstrated a nonlinear trend. This can be explained by the increase in the concentration polarization on the membrane surface, causing the increase in the osmotic pressure. The increase in the osmotic pressure thus requires the feed pressure to increase, and subsequently, the permeate flux increases proportionally with the increase in the system recovery ratio. This is one of the underlying operating dynamics of the enhancement in salt rejection witnessed. Figure 63 part (b) illustrates the system recovery ratio, permeate flux, and salt rejection relationship.

5.4.3. Influence of Feed Water Pressure on RO Membrane Performance

The theoretical results show that salt rejection and permeate flux are both rising in trend with a rise in the feed water pressure. The rise in permeate flux with increasing feed water pressure has a more or less linear relationship. In contrast, the salt rejection rate exhibits a sharp rise with feed pressures of up to 50 bar, following which the increase rate becomes less steep, ultimately plateauing at a feed pressure of around 70 bar. As displayed in Figure 63 part (c), the improvement in salt rejection is due to the higher net driving pressure, which also improves the permeate flux at the same time. The leveling of salt rejection at high feed pressures can be explained by the intrinsic limitation of the RO membrane as a barrier to feed water-dissolved salts; there is always a constant and measurable amount of salt permeation through the membrane. As feed pressure increases, the efficiency of these salt channels is improved since water is pushed against the membrane at a higher rate compared to the rate of salt transport. However, there is a critical point beyond which salt rejection cannot be further improved by exposing it to more feed pressure. This is echoed in the salt rejection curve shown in Figure 63 part (c), where it can be observed that after a particular pressure point, the rejection of salt does not keep on rising and there is also a residual flow of salt as well as the permeate flow across the membrane.

5.4.4. Effect of Feed Water Temperature on RO Membrane Performance

The simulation results recognize the permeate flux as being positively correlated with an increase in feed water temperature, with the salt rejection having a corresponding reduction under the same conditions. Further, an increase in feed water pressure has the effect of both increasing the permeate flux and reducing the salt rejection, with the former being at non-linear rates of increase. The decrease in salt rejection that is observed can be explained by the rise in the membrane permeability coefficient, where an increased quantity of salt ions is allowed to cross the membrane. Figure 63 part (d) shows the correlation between feed water temperature, permeate flux, and salt rejection.

5.5. RO Membrane Simulation Results

A significant benefit of utilizing a simulation model of RO system lies in its capability to determine the ideal range of operating feed pressures for application in a real plant. The operating pressure is defined by the upper limit of feed pressure, which is governed by the manufacturer's specifications concerning the physical characteristics of the membrane. However, the lowest feed pressure is identified by the osmotic pressure needed to prevent reverse membrane flow. The findings realized from the simulation provided in Figure 64 are representative of the highest and lowest applicable values of feed pressures in the selected RO plant taking into consideration that the feed water contains a salt concentration level of 35,000 ppm and a temperature of 22°C. Table 7 show the maximum, and minimum operating pressure and the corresponding recovery ratio and permeate flux values results from the simulation program.

Table 7. Corresponding simulation results for recovery ratio and permeate flux at maximum, and minimum operating pressure

	Pressure (bar)	Recovery ratio (%)	Permeate flux ($L/m^2/hr.$)
Max.	65	31	17.14
Min.	45	17	7.27

Theoretical analysis of the membrane showed that operating pressure is not altered with the fluctuations in water temperature. Nevertheless, as the temperature rises from 18°C to 26°C, there is an obvious recovery ratio, which ascends from 17% to 31%. In particular, an increase in temperature from 18°C to 26°C reduces the applied pressure from 63 bar to 56 bar for the case of water salinity rate of 30,000 ppm but still a flow rate of the permeate as 3.74 m^3/hr. Similarly, when the water salinity is 35,000 ppm at a permeate flow rate of 3.08 $m^3/hr.$, the operating pressure decreases from 63 bar to 57 bar, and when the salinity is 40,000 ppm at a flow rate of 2.53 $m^3/hr.$ for the permeate, the pressure decreases from 63 bar to 57.5 bar. Figure 65 presents the proposed operating charts with feed pressure as a function of salt rejection percentage vs. recovery ratio at various salinities and water temperatures.

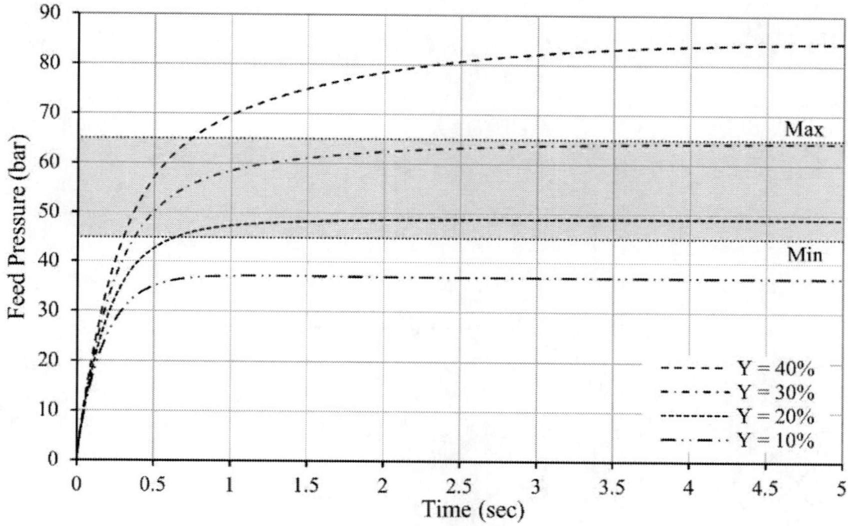

Figure 64. Maximum and minimum allowed operation pressure at feed water salinity of 35,000 ppm, and water temperature of 22°C.

Conclusion

The findings from model simulation and verification indicated that the presence of an accumulator in the system has a great impact on the piston pump pressure irregularity coefficient (εp). Further, the existence of a charged accumulator is able to suppress some of the detrimental effects linked with the malfunction of the piston pump.

Also, the permeability of the reverse RO membrane becomes the dominant influence on its operating pressure according to variations in feed water salinity, feed water temperature, or recovery ratio. To establish the appropriate operating pressure for the RO unit, both the physical properties of the membrane material and the osmotic pressure at the membrane exit concentration must be considered.

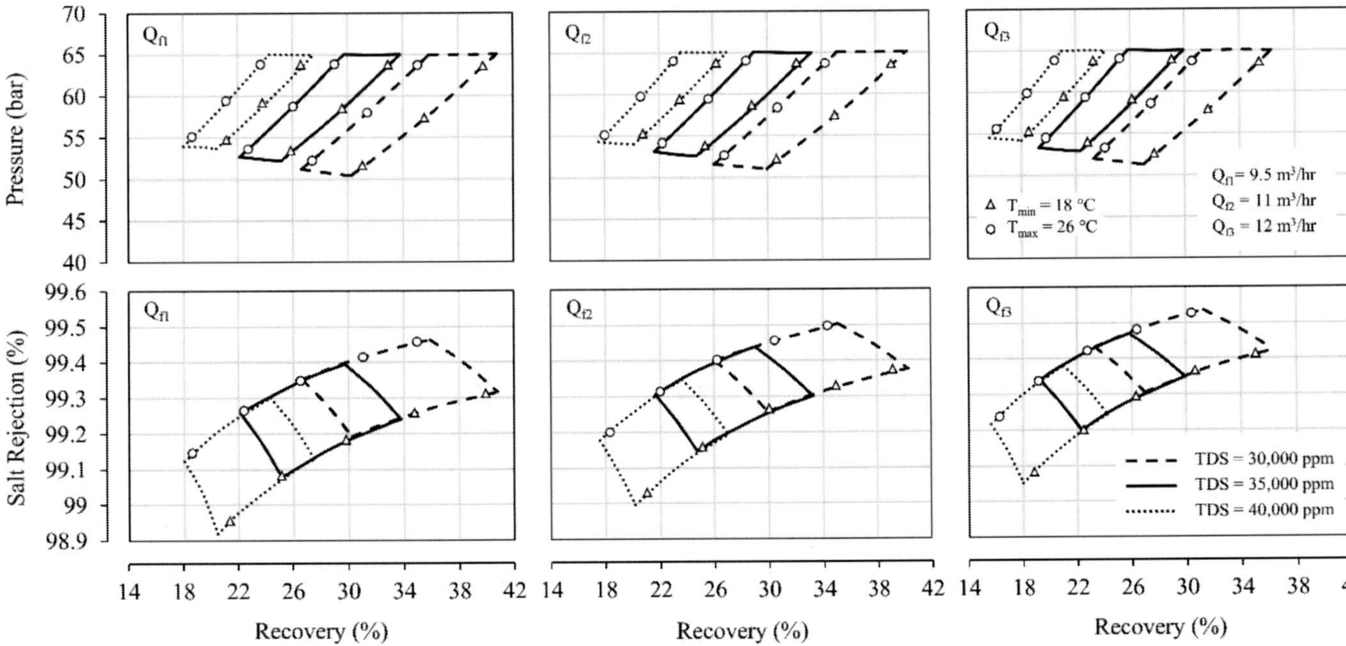

Figure 65. Suggested Operating Charts (Feed Pressure and % of Salt Rejection vs. Recovery Ratio) for Different Salinities at Different Feed Water Temperatures from 18°C to 26°C.

Theoretical analysis of the membrane showed that the influence of water temperature on operating pressure can be considered insignificant. Nevertheless, when the recovery ratio is raised from 17% to 31% together with temperature increment from 18ºC to 26ºC, there are great differences in operating pressure. Specifically, when temperature is raised from 18ºC to 26ºC, feed pressure decreases from 63 to 56, 57, and 57.5 bar at water salinity of 30,000, 35,000, and 40,000 ppm for a permeate flow rate of 3.74, 3.08, and 2.53 m^3/hr. in turn. The RO plant simulation software that was developed has been verified and proven to be appropriate for theoretical investigation which can result in significant cost and time savings. It enables the prediction of the effect of variations in different operating parameters such as salinity, water temperature, pump pressure or flow rate, and membrane type on the performance of the plant. It also enables the optimization of operating parameters and aids troubleshooting activities in the RO plant.

Chapter 9

Desalination Artificial Intelligence and Simulation Insights

Abstract

This chapter presents an overview of using advantageous Artificial Intelligence coupled with simulation techniques (AI-Sim) to optimize desalination processes and enhance operational efficiency. The AI-Sim concept covers several tools to face significant challenges in desalination systems, such as high operational costs, high energy requirements, maintenance costs, and water quality compliance. These tools analyze data to give predictive analytics and operational insights for system performance. Key focus areas include the AI-Sim operation and maintenance optimization tool, Water quality enhancement tool and total cost reduction and economics tool designed to optimize resource management and reduce operational costs. Additionally, operation staff skills enhancement tool is presented to suggest enhancement of the performance of operation staff through assessment and determination of training requirements. Also, operation staff skills enhancement tool is introduced to propose improvement in operation staff performance by way of evaluation and training needs determination. Lastly, the chapter presents overall equations of desalination plants and centrifugal pumps and research and development (R&D) issues to deal with performance indicators and sustainability. Overall outcomes of this chapter establish the innovative potential of AI-based analytics in designing more efficient, cost-advantaged, and sustainable desalination plants.

Keywords: desalination challenges, desalination performance optimization, operation and maintenance, water quality, economics, social challenges, artificial intelligence, machine learning

1. Introduction

Among the primary challenges for the desalination industry is the high cost of construction and operation of desalination plants and, consequently, the cost of water produced. Enhancing plant design and capacity, minimizing energy costs, and taking advantage of widespread technological development in membrane technologies and system designs can have a visible effect on the efficiency of operation and enhance water quality [49]. The principal uncertainty that is present in the operation and building of desalination plants, along with the inherently variable nature of climatic conditions, renders it extremely challenging to determine the long-term viability or efficiency of newly built desalination plants. Furthermore, although the price of desalinated water varies significantly, the performance of desalination plants can be greatly enhanced by considering the specific geographical and economic characteristics of countries that have these plants. Some of the most significant parameters that influence the overall cost of desalination include environmental and political limitations, the type of contracts associated with projects, the costs of labor and energy, plant capacity, salinity of the source water, the actual desalination techniques employed, and some other relatively insignificant factors [44].

Performance instability, induced by variations in the feed stream purity and high energy consumption, remains a bane despite concerted efforts at reducing cost. As such, interest is intensified in the application of artificial intelligence (AI) for improving process efficiency and cost-effectiveness of ocean water desalination [178]. Mathematical equations, although not necessarily able to capture the complexities of real situations, provide useful solutions by the combination of different simplifying assumptions. For reverse osmosis (RO) systems, mathematical modeling is used to enhance the design and operating characteristics by forecasting the effect of operating parameter changes such as feed water temperature, pressure, and salinity. The ensuing model equations allow for optimization of membrane selection and design, make possible predictions of the influence of fouling on system performance, and provide estimates of RO system life. Consequently, mathematical modeling of RO is an important tool for engineers and researchers involved in water treatment because it leads to improved insight into the optimization of RO system performance. The process of searching for the best solution to a specific problem by comparing different alternatives is referred to as optimization. Thus, developing a clear mathematical model is significant to keep the RO system at its optimum efficiency level, which can result in the reduction of total costs [179, 194].

In order to effectively model and represent the process of desalination, recent studies have suggested and investigated the application of AI techniques in combination with data-driven techniques. The classical model-based methods have been proven to be very difficult to correctly model the complexity of the desalination process, and they tend to be limited. The ability of AI-based models to transform massive data sets into relevant information and reliable behavioral models is the primary benefit that can be gained from using these techniques [110, 202].

One particular way in which AI has the potential to contribute to the optimization of desalination processes is by enabling the prediction of when membranes need to be cleaned or replaced. Over time, membranes may accumulate contaminants, causing their operating efficiency to decrease. By using Machine Learning (ML) algorithms to analyze data on water flow, pressure, and other factors, operators of desalination plants can predict when membranes are likely to become clogged and proactively replace them before they become a problem. This approach is known as predictive maintenance [203]. Furthermore, Al can help optimize desalination is by predicting water quality. The quality of seawater can vary depending on factors such as temperature, salinity, and weather conditions. By using ML algorithms to analyze data on water quality, desalination plants can adjust their operations to ensure that the water produced is consistently of high quality. This approach is known as predictive water quality [204]. Al can also help optimize energy usage in desalination plants. By analyzing data on energy consumption, ML algorithms can identify areas where energy usage can be reduced, such as by altering the velocity of water through the membranes or optimizing the use of pumps and other equipment [205]. Membrane separation, water purification, wastewater treatment, membrane separation, and water for agricultural systems are just some of the areas where Al and MI. have Conventional techniques of water purification include shortcomings that have been seen as opportunities for ML, which is grounded in mathematics, and AI, which has its foundation in the human intellect. AI and ML interventions can improve the efficiency and predictability of water desalination operations. Many different ML algorithms and those developed with Al [206].

In this chapter we will introduce insights examples for using desalination artificial intelligence, the main purpose of using desalination AI to overcome desalination challenges and achieve desalination sustainability. Using artificial intelligence in integration with simulation (AI-Sim), for desalination systems optimization, improvement, and development in many fields such as; economics improvement, technical performance, design, products and supplies quality, environmental impact, social impact, information and operation technology convergence, engineers and workers performance, ability to make decisions, applied scientific research, will be useful to achieve smart desalination systems.

Insights examples will be presented for using AI-Sim in water desalination in several frameworks, such as; operation and maintenance optimization, water quality enhancement, total cost reduction and economics, operation staff skills enhancement, practical enhancement, characteristic equations coefficients enhancement, research and development enhancement, process design optimization, minimize negative environmental impact, social impact enhancement, and information and operational technologies convergence, as shown in Figure 66. AI-Sim integration in desalination industry is requisite to secure present and future water supply.

The AI-Sim Tolls Framework for Desalination is a comprehensive platform that leverages artificial intelligence and simulation tools to optimize desalination processes. This framework aims to improve as example desalination plants' efficiency, sustainability, and cost-effectiveness to have smart desalination systems. Desalination AI-Sim tolls key components are data collection and integration, AI-powered predictive modeling, simulation tools, optimization algorithms, and real-time monitoring and control.

Data collection and integration by gathering data from various sources, including sensors, weather forecasts, and plant operations. AI-powered predictive modeling utilizes machine learning algorithms to predict water demand, optimize plant operations, and detect potential issues. Simulation tools by employing simulation software to model different desalination scenarios, including varying water sources, treatment processes, and energy inputs. Optimization algorithms through applying optimization techniques to identify the most efficient and cost-effective desalination strategies. Real-time monitoring and control by implementation real-time monitoring and control systems to ensure optimal plant performance.

Desalination AI-Sim tolls have a lot of benefits such as, improved efficiency, increased sustainability, enhanced reliability, and data-driven decision-making.

Improved efficiency through desalination process optimization to reduce energy consumption and costs. Increased sustainability by minimizing environmental impact by reducing waste and emissions. Enhanced reliability by predicting and preventing potential issues to ensure consistent water supply. Data-driven decision-making by providing insights and recommendations to inform desalination plant operations and maintenance. This framework can help desalination plants operate more efficiently, sustainably, and cost-effectively, ultimately providing a reliable source of clean water. A general Desalination AI-Sim toll framework shown in Figure 67.

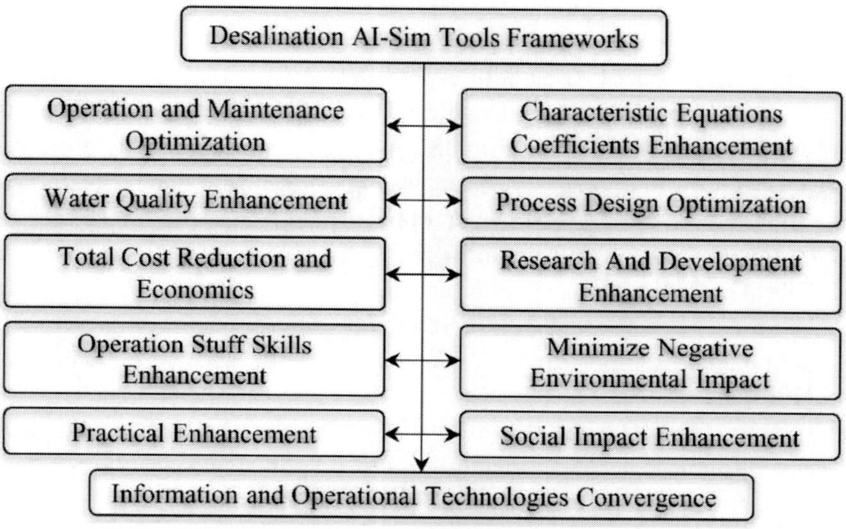

Figure 66. Desalination AI-Sim Tools Frameworks.

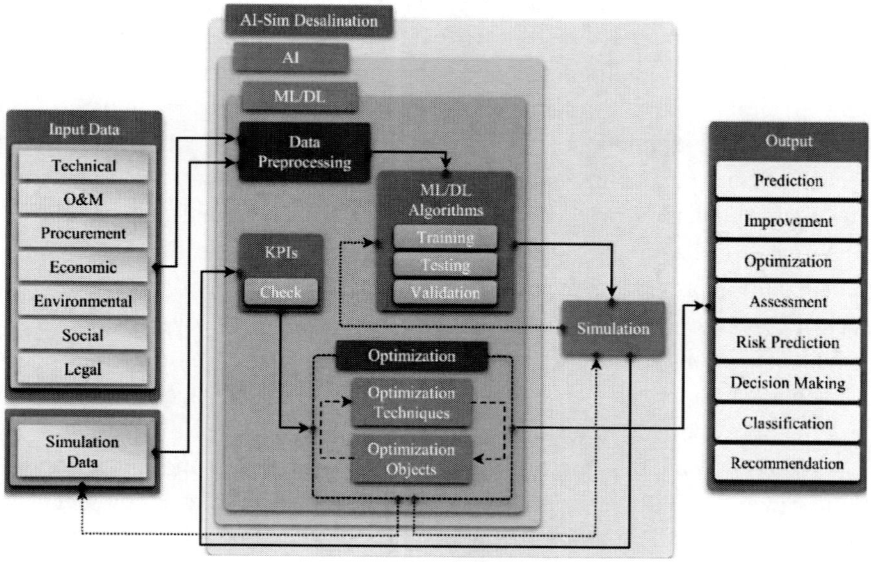

Figure 67. Desalination AI-Sim Tolls Frameworks.

2. Operation and Maintenance Optimization

Desalination systems face operation and maintenance challenges such as high energy consumption, high maintenance costs, and varying water quality. AI-Sim operation and maintenance optimization tool can optimize desalination system operation and maintenance, improving efficiency, reducing costs, and enhancing water quality. AI-Sim operation and maintenance optimization tool requires an input and output as follows.

Tool inputs, AI-Sim operation, and maintenance optimization tool data acquisition can be collected from seven main sources: technical data, O&M data, simulation data, procurement data, environmental data, social data, and economic data.

Technical data includes; design data for the desalination system, including all related studies, design drawings of the desalination system, technical specifications of the components and equipment of the desalination system, and design standards and codes. O&M data includes; real-time operational data of the desalination system under research and development, historical operation and maintenance data of the desalination system under research and development, real-time operational data of other desalination systems, historical operation and maintenance data of other desalination systems, and initial operation and maintenance plans. Simulation data include data generated from simulation and optimization processes. Procurement data includes; data on executing companies, and data on spare parts suppliers and consumable materials providers. Environmental data. Social data includes; social data of the plant's operational area, and operator data, including specialization, previous experience, educational background, and training level. Economic data including construction costs and operational cost elements, Tool outputs, AI-Sim operation and maintenance optimization tool in demand outputs are prediction, improvement, decision-making, classification, and recommendation [44, 110, 207-209].

Prediction includes; predicting failures in advance, and determining maintenance schedules according to the technical needs of the equipment and in alignment with the maintenance guidelines provided by the manufacturer. Improvement includes; proposing improvements for social impact, providing design improvement suggestions for designers, and Providing product improvement suggestions for manufacturers. Decision-making includes; analyzing operational data and assisting in decision-making, making critical decisions in predefined emergency situations to protect the desalination system from physical damage, and identifying spare parts and consumables (e.g., chemicals, oils, grease, media, membranes, etc.) for an upcoming period and providing procurement recommendations along with suggested specifications based on actual operational needs. Classification includes, classifying consumable materials (e.g., chemicals and membranes) according to their technical quality and economic value under actual operating conditions and their impact on desalination system performance indicators. Recommendation includes; developing operational and maintenance scenarios to minimize downtime, aligning future operations based on actual and expected consumption, proposing strategies to reduce production costs (operation and maintenance), and Suggesting measures to minimize negative environmental impact. As shown in Figure 68.

AI-Sim operation and maintenance optimization framework consists of six layers, data integration layer, AI/ML and simulation layer, optimization layer, decision support system (DSS) layer, maintenance and update layer and cybersecurity layer.

Data integration layer includes data collection, data preprocessing, and data storage. Data collection involves gathering information from diverse sources such as sensors, control systems, and maintenance management systems. Data preprocessing is necessary to guarantee the quality, accuracy, and consistency of the data. Data storage to store data in a secure and scalable repository. AI/ML and Simulation layer includes AI, simulation models, and digital twins. AI algorithms, such as machine learning and deep learning, are utilized to analyze data and optimize operations and maintenance. Developing simulation models of desalination systems to predict behavior and optimize operation and maintenance [110]. Creating a digital twin of the desalination system to simulate and optimize operation and maintenance.

Optimization layer includes optimization algorithms, multi-objective optimization, and real-time optimization. Optimization algorithms, including linear programming and genetic algorithms, are utilized to enhance operation and maintenance processes. The process involves the optimization of multiple objectives, including energy consumption, water quality, and maintenance costs. The system optimizes operations and maintenance in real-time, adapting to changes in water demand, energy prices, and other relevant factors. DSS layer includes visualization, alerts and notifications, and recommendations [210]. The DSS layer facilitates decision-making by visualizing data and optimization results. The system also provides alerts and notifications for abnormal conditions and maintenance needs.

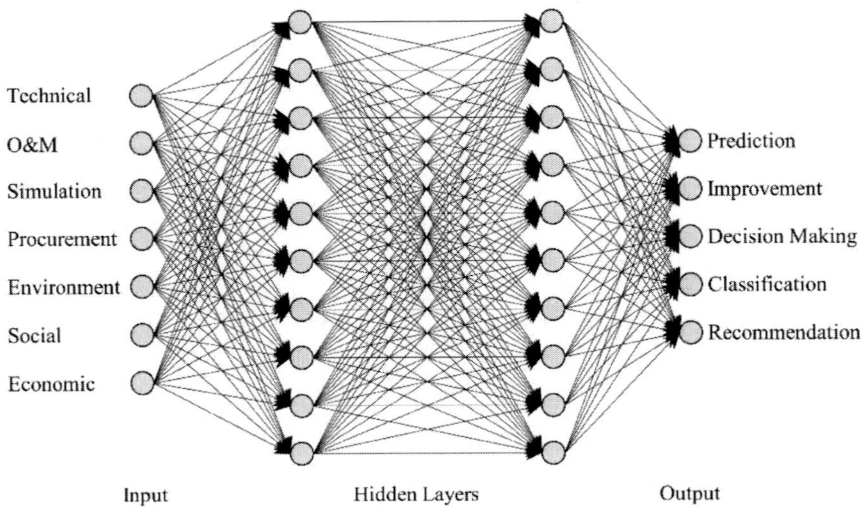

Figure 68. Architecture of AI-Sim Operation and Maintenance Optimization Tool Model.

The system also provides recommendations for optimizing operation and maintenance processes. Maintenance and update layer includes maintenance scheduling, software updates, training, and support. Scheduling maintenance activities based on optimization results. Updating software and models to ensure accuracy and effectiveness. Providing training and support for users to ensure effective use of the framework. Cybersecurity layer includes data encryption, access control, and intrusion detection. The encryption of data ensures its confidentiality and integrity, while the control of access to the framework ensures its authorized use. Detecting and responding to cyber threats to ensure the integrity of the framework [211].

AI-Sim operation and maintenance optimization tool can be used for desalination operation optimization and maintenance optimization in areas such as: Energy Consumption Optimization, Plant Availability Optimization, Real-Time Optimization, Predictive Maintenance, Maintenance Scheduling Optimization and Spares Optimization.

Energy Consumption Optimization to optimize energy consumption, reducing energy costs and environmental impact. Water Quality Optimization aims to enhance water quality, guaranteeing that it either meets or surpasses regulatory standards. Plant Availability Optimization aims to optimize plant availability, thereby reducing downtime and increasing productivity. Real-Time Optimization optimizes the operation of the desalination system in real-time, adapting to fluctuations in water demand, energy prices, and other relevant factors. Predictive Maintenance to predict maintenance needs, reducing unplanned downtime and increasing productivity. Maintenance Scheduling Optimization aims to optimize maintenance scheduling, thereby reducing maintenance costs and enhancing plant availability. Spares Optimization optimizing spares inventory, reducing inventory costs, increasing plant availability, and Root Cause Analysis is a crucial process that identifies the underlying causes of equipment failures and other issues. AI/ML algorithms for AI-Sim operation and maintenance optimization tools include ML Algorithms, Deep Learning (DL) Algorithms, Optimization Algorithms, Anomaly Detection Algorithms and Predictive Maintenance Algorithms.

ML algorithms include Linear Regression, Decision Trees, Random Forest, and Support Vector Machines (SVMs) [212]. Linear regression predicts continuous outcomes like energy consumption or water quality, decision trees classify equipment conditions or predict maintenance needs, random forests use ensemble learning to predict equipment conditions or maintenance needs, and SVMs classify equipment conditions or predict maintenance needs using kernel methods. DL Algorithms include Convolutional Neural Networks (CNNs), Recurrent Neural Networks (RNNs), and Long Short-Term Memory (LSTM) Networks. CNNs look at sensor data, like images or time-series data, to guess what kind of maintenance or repair work needs to be done on equipment. RNNs guess what kind of maintenance or repair work needs to be done on equipment by using sequential data, like time-series data. LSTM Networks guess what kind of maintenance or repair work needs to be done by using sequential data, like time-series data [213].

Optimization Algorithms include Linear Programming, Non-Linear Programming, Genetic algorithms, and Particle Swarm Optimization [214, 215]. Linear Programming optimizes operation and maintenance schedules using linear objective functions and constraints, Non-Linear Programming optimizes operation and maintenance schedules using non-linear objective functions and constraints, Genetic Algorithms optimizes operation and maintenance schedules using evolutionary algorithms, and Particle Swarm Optimization optimizes operation and maintenance schedules using swarm intelligence algorithms. Anomaly Detection Algorithms include One-Class SVM, Local Outlier Factor (LOF), and Isolation Forest. The One-Class SVM uses kernel methods to find outliers in equipment conditions or maintenance needs. The LOF uses density-based methods to find outliers in equipment conditions or maintenance needs. The Isolation Forest uses ensemble methods to find outliers in equipment conditions or maintenance needs [216].

Predictive Maintenance Algorithms include Prognostic Models, Fault Detection and Diagnosis and Condition-Based Maintenance [217]. Prognostic Models to predict equipment's remaining useful life using machine learning and statistical methods, Fault Detection and Diagnosis to detect and diagnose equipment faults using machine learning and statistical methods, and Condition-Based Maintenance to schedule maintenance based on equipment condition using machine learning and statistical methods. Simulation modules for AI-Sim operation and maintenance optimization tool includes Desalination Process Simulation, Energy Consumption, Maintenance and Reliability, Water Quality, and Economic and Financial.

Desalination Process Simulation includes for example RO Simulation, Multi-Stage Flash (MSF) Distillation Simulation and Multi-Effect Distillation (MED) Simulation. Simulating RO process performance, including membrane fouling, scaling, and permeate quality; simulating MSF distillation process performance, including heat transfer, mass transfer, and distillate quality; and simulating MED process performance, including heat transfer, mass transfer, and distillate quality. Energy Consumption simulation includes, for example, pumping system simulation and electrical power consumption simulation. The simulation encompasses the simulation of pumping system energy consumption, which includes pump performance, pipeline hydraulics, and control systems, as well as the simulation of electrical power consumption, which includes motor efficiency, transformer efficiency, and electrical distribution system losses. Maintenance and Reliability Simulation includes Equipment Failure Simulation, Maintenance Scheduling Simulation, and Reliability-Centered Maintenance (RCM) Simulation. The equipment failure simulation simulates equipment failure rates, including the mean time between failures (MTBF) and the mean time to repair (MTTR). The maintenance scheduling simulation simulates maintenance scheduling, which includes preventive, corrective, and condition-based maintenance. The RCM simulates RCM analysis, which includes failure modes and effects analysis and reliability block diagrams.

Water Quality Simulation includes; for example, water Simulation and Remineralization Simulation. Water Quality Modeling to simulate water quality parameters, including pH, temperature, turbidity, and salinity, The Disinfection Simulation simulates various disinfection processes such as chlorine, ozone, and UV disinfection, while the Remineralization Simulation mimics remineralization processes such as limestone contactors and chemical dosing systems. Operation and maintenance, and water quality are highly interrelated aspects in the case of desalination systems. The process design and operation are strongly dependent on the water quality standards required and the quality characteristics of feed water. Optimizing the system performance and meeting the regulatory requirements necessitates an understanding of these interrelations.

Economic and Financial Simulation includes Cost-Benefit Analysis Simulation, Life Cycle Cost (LCC) Simulation, and Return on Investment (ROI) Simulation. The Cost-Benefit Analysis Simulation simulates cost-benefit analysis, which includes capital costs, operating costs, and revenue projections; the LCC Simulation simulates LCC analysis, which includes equipment costs, maintenance costs, and energy costs; and the ROI Simulation simulates ROI analysis, which includes payback period, net present value (NPV), and internal rate of return. Simulation techniques include Discrete Event Simulation (DES), Continuous Simulation, Hybrid Simulation, Monte Carlo (MC) Simulation, Agent-Based Simulation, System Dynamics (SD) Simulation and Optimization Techniques [218, 219].

DES includes simulating processes, such as manufacturing, logistics, and supply chain management, and simulating systems, such as transportation, communication, and energy systems [220]. Continuous Simulation includes simulating dynamic systems, such as mechanical, electrical, and thermal systems, and simulating steady-state systems, such as chemical processes and power plants [221]. Hybrid Simulation includes combining DES and continuous simulation to model complex systems and combining simulation and optimization to optimize system performance [222]. MC Simulation includes risk analysis to analyze risk and uncertainty in systems, such as financial, engineering, and environmental systems, and sensitivity Analysis to analyze the sensitivity of system performance to changes in input parameters [223]. Agent-Based Simulation includes modeling complex systems, such as social networks, economies, and ecosystems, and analyzing emergent behavior in complex systems [224]. SD Simulation includes modeling feedback loops and dynamics in systems, such as population growth and resource depletion, and analyzing system behavior over time, including growth, decline, and oscillation [225].

Optimization Techniques include linear programming to optimize linear systems, such as resource allocation and production planning; non-linear programming to optimize non-linear systems, such as chemical processes and mechanical systems; genetic algorithms to optimize complex systems using evolutionary algorithms; and particle swarm optimization to optimize complex systems using swarm intelligence algorithms.

AI-Sim operation and maintenance optimization tool offers a powerful tool to improved efficiency, cost savings, enhanced reliability, improved water quality, increased safety, regulatory compliance and data-driven decision-making.

Improved efficiency includes optimizing operation and maintenance schedules to reduce downtime and increase productivity, reducing energy consumption through optimized operation and maintenance, and reducing water consumption through optimized operation and maintenance.

Cost savings include reducing maintenance costs through predictive maintenance and optimized maintenance scheduling, reducing energy costs through optimized operation and maintenance, and reducing water costs through optimized operation and maintenance. Enhanced reliability includes predicting equipment failures and performing maintenance before breakdowns occur, reliability-centered maintenance to optimize maintenance schedules based on equipment reliability and criticality, and reducing downtime through optimized operation and maintenance. Improved Water Quality includes optimizing water treatment processes to improve water quality, monitoring water quality in real-time to detect any changes or anomalies and automatically adjusting water treatment processes to maintain optimal water quality. Increased Safety includes predictive maintenance through predicting equipment failures and performing maintenance before failures occur, reducing the risk of accidents; automated safety protocols through automatically implementing safety protocols in response to changes in equipment condition or water quality; and real-time monitoring through monitoring equipment condition and water quality in real-time to detect any changes or anomalies. Regulatory compliance includes automatically generating reports to meet regulatory requirements, monitoring equipment conditions and water quality to ensure compliance with regulatory requirements, and providing alerts and notifications in response to changes in equipment conditions or water quality that may impact regulatory compliance. Data-driven decision-making real-time data analytics to support decision-making, providing predictive analytics to forecast future trends and patterns, and automatically providing decision support recommendations based on data analytics and predictive analytics.

Key Performance Indicators (KPIs) used for operation and maintenance optimization such as operational efficiency, maintenance efficiency, energy and water efficiency, reliability and uptime, cost savings, water quality, safety and compliance, [226, 227] system performance index, and environmental performance index.

Operational Efficiency includes overall equipment effectiveness to measure the percentage of manufacturing time that is productive, availability to measure the percentage of time that equipment is available for operation, and utilization to measure the percentage of time that equipment is actually operating [228]. Maintenance efficiency includes MTBF to measure the average time between equipment failures, MTTR to measure the average time it takes to repair equipment after a failure, and maintenance backlog to measure the number of outstanding maintenance tasks [229].

Energy and Water Efficiency include energy consumption to measure the amount of energy consumed by equipment and processes, water consumption to measure the amount of water consumed by equipment and processes, and specific energy consumption to measure the energy consumption per unit of production [230]. Reliability and uptime include dependability to measure the ability of equipment to operate without failure, uptime to measure the percentage of time that equipment is operating, and downtime to measure the percentage of time that equipment is not operating. Cost savings include maintenance cost savings to measure the reduction in maintenance costs achieved through optimized maintenance scheduling, energy cost savings to measure the reduction in energy costs achieved through optimized energy consumption, and water cost savings to measure the reduction in water costs achieved through optimized water consumption. Water quality includes Water Quality Index (WQI) to measure the overall quality of the water produced, pH Levels to measure the acidity or alkalinity of the water produced, and turbidity to measure the clarity or cloudiness of the water produced [231]. Safety and compliance include safety incidents to measure the number of safety incidents reported, compliance to measure the degree of compliance with regulatory requirements, and audit results to measure the results of audits conducted to assess compliance.

3. Water Quality Enhancement

Water quality enhancement is critical to ensure the produced water from the desalination system meets the required standards for drinking, industrial, or agricultural purposes. AI-Sim technologies can enhance water quality by optimizing maintenance schedules, streamlining operations, and enhancing water quality monitoring. Meanwhile, AI-Sim has to be used to analyze real-time data to guarantee stable water quality by modifying treatment parameters. Using AI-Sim applications for monitoring and modeling allows early detection of water contaminants and prediction of water quality changes through trend identification. AI-Sim applications support compliance with environmental regulations. AI-Sim water quality enhancement tool has an input and output as follows:

Tool inputs, AI-Sim water quality enhancement tool data acquisition can be collected from seven main sources: technical data, O&M data, simulation data, procurement data, environmental data, social data, and economic data. Technical data includes; design data for the desalination system, including all related studies, design drawings of the desalination system, technical specifications of the components and equipment of the desalination system, design standards and codes, chemical, physical, and biological specifications of raw water, and chemical, physical, and biological specifications of produced water according to international and local standards and contractual requirements. O&M data includes; real-time operational data of the desalination system under research and development, historical operation and maintenance data of the desalination system under research and development, real-time operational data of other desalination systems, historical operation and maintenance data of other desalination systems, and initial operation and maintenance plans. Simulation Data includes; data generated from simulation and optimization processes. Procurement data includes; data on executing companies, and data on spare parts suppliers and consumable materials providers. Environmental data. Social data includes; social data of the plant's operational area, and operator data, including specialization, previous experience, educational background, and training level. Economic data, including construction costs and operational cost elements.

Tool outputs, AI-Sim water quality enhancement tool in demand outputs are prediction, improvement, decision-making, classification, and recommendation.

Prediction includes; predicting future probabilities of non-compliance with water quality standards, identifying the causes, and proposing measures to prevent it. Improvement includes; reviewing the compliance of produced water quality with international and local standards, developing scenarios for adjusting operational processes to maintain water quality under various expected changes, especially raw water properties variations, providing design improvement suggestions for designers, and providing product improvement suggestions for manufacturers. Decision-making includes; analyzing operational data and assisting in decision-making. Classification includes;

classifying consumable materials (e.g., chemicals and membranes) based on their technical quality and economic under actual operating conditions and their impact on water quality performance indicators. Recommendation includes; proposing strategies for handling non-compliant produced water, considering health, environmental, and economic impacts. As shown in Figure 69.

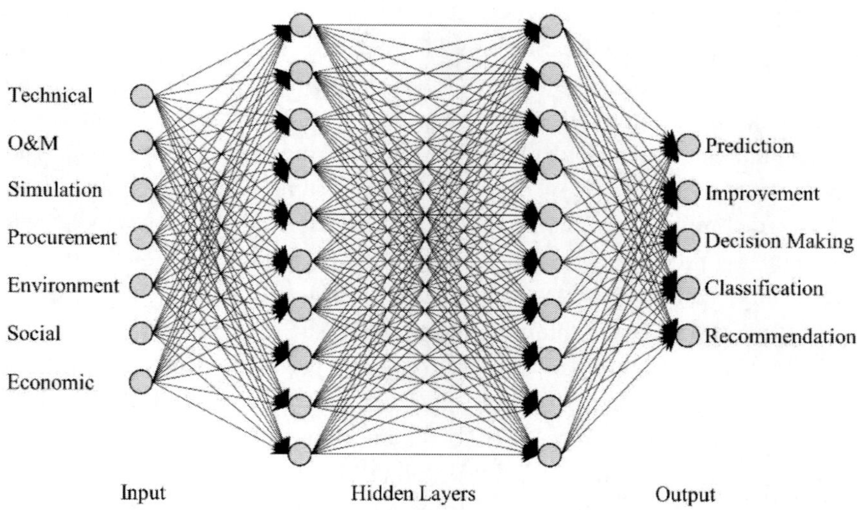

Figure 69. Architecture of AI-Sim Water Quality Enhancement Tool Model.

AI-Sim water quality enhancement framework consists of Data collection, Data preprocessing, Machine learning model, Optimization algorithm, Simulation and optimization, and Implementation and deployment.

Data collection is obtained from various sources, including instrumentation measurements from existing systems, equipment data, maintenance data, water quality data, and simulated data. Measured data to be collected through installed instrumentation in the desalination system, such as pressure, flow rate, temperature, and conductivity; equipment data to be collected from the installed equipment, such as pumps performance, motors performance, valve performance, membrane performance, and dosing systems performance; maintenance data to be collected from maintenance activities performed on the desalination system, such as maintenance schedules, repair records, replacement records, and chemical consumption records; water quality data to be collect from the water quality of the feedwater, permeate, and concentrate streams, such as pH, temperature, conductivity, turbidity,

COD, BOD, TSS … etc. Data preprocessing of collected data through three steps: data cleaning, data normalization, and data transformation. Data Cleaning to Clean the data by removing noise, handling missing values, and removing duplicates; Data Normalization to Normalize the data to a common scale to prevent feature dominance; Data Transformation Transform the data into a suitable format for machine learning algorithms, such as, Time-series data, Frequency-domain data, and Statistical data. Machine learning models are needed for desalination systems water quality enhancement, such as predictive models, classification models, and regression models. Predictive models were developed to forecast water quality based on operational parameters, such as feedwater temperature, feedwater salinity, membrane fouling, and permeate flow rate; classification models were developed to classify water quality based on operational parameters, such as good/bad water quality and compliant/non-compliant with regulations; and regression models developed to predict continuous water quality parameters, such as pH, conductivity, turbidity, etc. Optimization algorithms are needed for desalination system water quality enhancement, such as linear programming, non-linear programming, genetic algorithms, and particle swarm optimization, to optimize desalination system operation for enhanced water quality. Simulation and Optimization needed for desalination system water quality enhancement through Simulate the desalination system operation using the machine learning model and optimization algorithm to optimize the desalination system operation for enhanced water quality using the optimization algorithm, then evaluate the performance of the optimized desalination system operation using metrics such as; Water quality, Energy consumption, and Maintenance costs. Implementation and deployment are needed for desalination system water quality enhancement through Implementation: Implement the AI-Sim framework in a cloud-based or on-premise environment. Deploy the AI-Sim framework in the desalination system, integrating it with existing systems and infrastructure, and Monitor and maintain the AI-Sim framework to ensure optimal performance and accuracy.

AI-Sim water quality enhancement tool can be used for desalination system water quality enhancement in the areas such as predictive modeling, optimization of desalination processes, water quality monitoring and control, energy efficiency and cost reduction, environmental impact mitigation, regulatory compliance and reporting, and asset management and maintenance.

Predictive modeling to predict water quality parameters and forecast water demand. Predict water quality parameters such as salinity, pH, temperature, and turbidity, and forecast water demand to optimize desalination plant operations. Optimization of desalination processes to improve water quality and reduce energy consumption and minimize environmental impact. Water quality monitoring and control through real-time water quality monitoring and automated control systems. Monitor water quality in real-time to detect anomalies and predict water quality issues, and implement automated control systems to adjust desalination plant operations in response to changing water quality conditions. Energy efficiency and cost reduction to optimize energy consumption and the process involves identifying strategies for cost reduction. Optimize energy consumption to reduce costs and minimize environmental impact; identify cost reduction strategies through predictive maintenance, optimized operations, and supply chain optimization. Environmental impact mitigation to assess environmental impact and identify mitigation strategies. Assess the environmental impact of desalination plant operations and identify mitigation strategies to minimize environmental impact, such as optimizing intake and outfall systems. Regulatory compliance and reporting to ensure compliance with regulations and generate automated reports. ensure compliance with regulatory requirements and standards and generate automated reports to facilitate compliance reporting and stakeholder engagement. Asset management and maintenance to implement predictive maintenance strategies and optimize asset performance. Implement predictive maintenance strategies to minimize downtime, extend asset lifespans, and optimize asset performance to improve efficiency and reduce costs.

AI/ML algorithms for AI-Sim water quality enhancement tool can utilize a range of AI/ML algorithms, including ML Algorithms, DL Algorithms, Natural Language Processing (NLP) Algorithms, Optimization Algorithms, Time-Series Analysis Algorithms, and Ensemble Methods.

ML algorithms include Linear Regression, Decision Trees, Random Forest, SVMs, and Neural Networks. There are different ways to predict water quality parameters. Linear regression predicts continuous parameters, decision trees predict continuous parameters, random forest predicts continuous parameters, SVMs predict SVMs, and neural networks use complex patterns to predict water quality parameters. DL algorithms include CNNs, RNNs, and LSTM Networks. CNNs analyze water quality images or sensor data, RNNs predict water quality parameters using time-series data, and LSTM networks of neurons predict water quality parameters using time-series data. NLP algorithms include Text Classification and Sentiment Analysis. Text classification is used to categorize water quality reports or sensor data, while sentiment analysis analyzes these reports or data to determine their sentiment

[232]. Optimization algorithms include genetic algorithms, particle swarm optimization, and simulated annealing. Using genetic algorithms, particle swarm optimization, and probabilistic optimization can all be used to improve water treatment processes. These methods are based on scientific ideas about how things evolve. Time-series analysis algorithms include autoregressive integrated moving average and exponential smoothing. autoregressive integrated moving average to predict water quality parameters using time-series data and exponential smoothing to predict water quality parameters using time-series data [233]. ensemble methods include bagging, Boosting, and Stacking. The aim is to enhance prediction accuracy through the integration of multiple models.

For improving the water quality in a desalination system, simulation modules are needed. These include modeling for the pretreatment process, the RO process, and the post-treatment process; membrane fouling; modeling for water quality; and sensitivity analysis. Some of the things that can be modeled are membrane fouling modeling, simulation, and prediction; water quality modeling to look at, simulate, and predict things like pH, turbidity, Total Dissolved Solids (TDS), and more; dynamic modeling for time-dependent simulations; and sensitivity analysis to check how sensitive parameters affect water quality.

Simulation modules for AI-Sim water quality enhancement tool can include the following simulation modules, Water Quality Prediction Module, Desalination Process Simulation Module, Water Treatment Process Simulation Module, Hydraulic and Hydrodynamic Simulation Module, Energy Efficiency and Cost Simulation Module, and Environmental Impact Simulation Module.

Water quality prediction module include physical-chemical modeling, biological modeling, and ML modeling. The module uses physical-chemical modeling to simulate physical-chemical processes that affect water quality, biological modeling to simulate biological processes that affect water quality, and ML modeling to predict water quality parameters using ML algorithms [234]. Desalination process simulation module includes RO Simulation, MSF Simulation, and MED Simulation RO processes Simulations include, for example, membrane fouling and scaling, MSF processes, Simulations include, as examples, heat transfer and mass balance and MED processes. Simulations include, for examples, heat transfer and mass balance. Water treatment process simulation module includes for example coagulation and flocculation simulation, sedimentation and flotation simulation, and filtration simulation. The simulation of coagulation and flocculation processes includes how particles settle and how well they are removed; the simulation of sedimentation and flotation processes includes how particles settle and how well they are removed; and the simulation of filtration processes includes how well particles are removed and how often filters get clogged. Hydraulic and hydrodynamic simulation module include for example, pipe network simulation, tank and reservoir simulation, and pump station simulation. The pipe network simulation simulates the hydraulics of the pipe network, including pressure, flow rate, and water quality; the tank and reservoir simulation includes water level, pressure, and water quality; and the pump station simulation replicates the operations of the pump station, including pump performance, energy consumption, and water quality. Energy efficiency and cost simulation module include energy consumption simulation and cost simulation. The module includes a simulation of energy consumption for desalination and water treatment processes, as well as a cost simulation that accounts for energy, chemicals, and maintenance associated with these processes. Environmental impact simulation module includes intake and outfall simulation, chemical usage simulation, and greenhouse gas emissions simulation. The intake and outfall simulation model the effects of intake and outfall systems on marine life and water quality. The chemical usage simulation models the effects of using chemicals in desalination and water treatment processes on the environment. Finally, the greenhouse gas emissions simulation models the effects of desalination and water treatment processes. Simulation techniques include DES, SD Simulation, Agent-Based Modeling (ABM) Simulation, Cellular Automata (CA) Simulation, MC Simulation, Dynamic Simulation, Stochastic Simulation, and Hybrid Simulation.

DES includes modeling of individual events and queueing theory. DES models' individual events, such as water intake, treatment, and distribution, while queueing theory models waiting times and queues in treatment processes. SD simulation includes stock and flow modeling and feedback loop analysis. The stock and flow modeling simulates the accumulation and flow of water quality parameters, while the feedback loop analysis examines the feedback loops in water treatment systems. ABM simulation includes agent-based modeling and emergent behavior analysis. ABM simulates the behavior of individual agents, such as water molecules or microorganisms, while emergent behavior analysis analyzes the emergent behavior of complex systems. CA simulation includes grid-based modeling and local interaction analysis. The grid-based modeling method simulates water quality parameters on a grid-based system, while the local interaction analysis method analyzes the local interactions between neighboring cells. MC simulation includes random sampling and probability distribution analysis. Random sampling to simulate uncertainty in water quality parameters using random sampling and probability distribution analysis to analyze probability distributions of water quality parameters. Dynamic simulation includes time-series analysis and control system analysis. Time-series analysis simulates the behavior of water quality parameters over time, while control system analysis analyzes the control systems in water treatment processes. Stochastic simulation includes randomized process modeling and uncertainty analysis. Random process modeling: to simulate random processes in water quality parameters, and uncertainty analysis to analyze uncertainty in water quality parameters [235]. Hybrid simulation includes a combination of simulation techniques and integration of different modeling paradigms. Hybrid simulation combines multiple simulation techniques to simulate complex systems and integrates different modeling paradigms to further simulate complex systems.

AI-Sim water quality enhancement tool would offer a powerful tool to Predict and Prevent Water Quality Issues, Optimize Water Treatment Processes, Improve Water Quality and Safety, Enhance Decision-Making and Planning, Reduce Costs and Environmental Impact, Improve Regulatory Compliance and Reporting, Enhance Public Health and Safety.

Predict and prevent water quality issues include predictive modeling and early warning systems. Predictive modeling to predict water quality parameters and identify potential issues, and early warning systems to provide early warnings for potential water quality issues. Optimize water treatment processes including process optimization and energy and chemical usage reduction. Process optimization to optimize water treatment processes to improve efficiency and effectiveness and energy and chemical usage reduction to reduce energy and chemical usage in water treatment processes. Improve water quality and safety include water quality monitoring and contaminant detection and removal. Water quality monitoring to monitor water quality parameters in real-time, and contaminant detection and removal to detect and remove contaminants from water supplies. Enhance decision-making and planning include data-driven decision-making, and scenario planning and analysis. The use of data-driven decision-making not only provides valuable insights for decision-making, but also facilitates the planning and analysis of future water quality challenges. Reduce costs and environmental impact include cost reduction and environmental impact reduction. Cost reduction to reduce costs associated with water treatment and maintenance and environmental impact reduction to reduce the environmental impact of water treatment processes. Improve regulatory compliance and reporting include regulatory compliance and automated reporting. Regulatory compliance ensures adherence to regulatory requirements, while automated reporting generates reports for regulatory agencies and stakeholders. Enhance public health and safety include waterborne disease prevention and improved water quality. Waterborne disease prevention to prevent waterborne diseases through early detection and removal of contaminants and improved water quality to provide safe and clean drinking water for communities.

AI-Sim water quality enhancement offers a powerful tool to improve water quality, increase recovery, reduce TDS, reduce energy consumption, extend membrane lifetime, comply with regulatory standards, and enhance decision-making. However, there are some challenges, including data quality, which is crucial for accurate predictions and optimization; model complexity, which can impact computational efficiency and accuracy; and scalability, which is crucial for large-scale RO desalination plants. AI-Sim water quality enhancement tool can provide a robust and efficient solution for water quality enhancement in desalination systems.

AI-Sim's water quality enhancement optimization objective is to minimize TDS, maximize water recovery, optimize pH levels, minimize contaminant concentrations, and improve the WQI. Optimization techniques to be used for water quality enhancement include genetic algorithms, particle swarm optimization, simulated annealing, dynamic programming, linear programming, and non-linear programming. KPIs used to evaluate water quality enhancement such as TDS, Recovery, pH, Turbidity, Contaminant Concentrations, WQI, chemical and physical characteristics, and Specific Energy Consumption.

KPIs are used for water quality enhancement such as Water Quality KPIs, Treatment Efficiency KPIs, Compliance and Regulatory KPIs, Operational and Maintenance KPIs, Financial and Economic KPIs, and Environmental and Sustainability KPIs.

Water quality KPIs such as pH levels, Turbidity, TDS, Bacterial contamination, and Chemical contamination. pH levels to measure the acidity/basicity of water, turbidity to measure the clarity of water, TDS to measure the concentration of dissolved solids in water, bacterial contamination to measure the presence of bacteria, such as E. coli, and chemical contamination to measure the presence of chemicals, such as heavy metals [226]. Treatment efficiency KPIs such as Removal efficiency, Treatment capacity, and Energy consumption. Removal efficiency measures the percentage of contaminants removed during treatment, treatment capacity measures the volume of water treated per unit of time, and energy consumption measures the energy used per unit of water treated. Compliance and regulatory KPIs such as regulatory compliance, Permit compliance, and Reporting accuracy. The regulatory compliance KPI measures adherence to regulatory requirements, the permit compliance KPI measures compliance with permit conditions, and the reporting accuracy KPI measures the accuracy of reporting to regulatory agencies. Operational and maintenance KPIs such as Uptime and availability, maintenance frequency, and MTTR [236]. Uptime and availability measure the percentage of time the treatment plant is operational; maintenance frequency measures the frequency of maintenance activities; and MTTR measures the time taken to repair equipment. Financial and economic KPIs such as cost per unit of water treated, ROI and cost savings. Cost per unit of water treated to measure the cost of treating one unit of water, ROI to measure the financial return on investment in water treatment infrastructure, and cost savings to measure the cost savings achieved through efficient treatment processes. Environmental and sustainability KPIs such as Carbon footprint, Water reuse and recycling, and environmental impact. Carbon footprint to measure the greenhouse gas emissions associated with water treatment, water reuse, and recycling to measure the percentage of treated water reused or recycled and environmental impact to measure the impact of water treatment on the environment. These KPIs provide a comprehensive framework for measuring the effectiveness of water quality enhancement initiatives.

4. Total Cost Reduction and Economics

The framework encompasses water consumption, water source, and economic resource management. AI-Sim-powered predictive analytics will help monitor and forecast water consumption needs and availability, enabling more techno-economically efficient water resource management. AI-Sim Total Cost Reduction and Economics tool has an input and output as follows:

Tool inputs, AI-Sim total cost reduction and economics tool data acquisition can be collected from seven main sources: technical data, O&M data, simulation data, procurement data, environmental data, social data, and economic data.

Technical Data includes design data for the desalination system, including all related studies, design drawings of the desalination system, technical specifications of the components and equipment of the desalination systems, and design standards and codes. O&M data includes real-time operational data of desalination systems under research and development, historical operation and maintenance data of desalination system under research and development, real-time operational data of other desalination systems, historical operation and maintenance data of other desalination systems, and initial operation and maintenance plans. Simulation data includes data generated from simulation and optimization processes. Procurement data includes data on executing companies, and data on spare parts suppliers and consumable materials providers. Environmental data. Social data includes social data of plant's operational area, and operator data, including specialization, previous experience, educational background, and training level. Economic including construction costs and operational cost elements [237].

Tool outputs, AI-Sim total cost reduction and economics tool in demand outputs are improvement, assessment, optimization, decision-making, and recommendation. Improvement includes linking operation and maintenance with production needs to reduce costs, and developing scenarios for operation and maintenance to minimize downtime, considering maintenance scheduling based on equipment technical needs and manufacturer guidelines, while consolidating possible maintenance tasks to reduce overall plant downtime. Assessment includes identifying energy losses (e.g., operating equipment in low-efficiency performance zones), identifying consumables losses (e.g., excessive use of chemicals, premature or delayed membrane replacements), assessing the impact of design efficiency on production costs (construction/operation), assessing the impact of operator efficiency on operational costs, and assessing the impact of consumable quality on operational costs. Optimization includes reviewing economic performance indicators and their impact on production costs and return on investment and reviewing the comparison between production costs and pricing tariffs. Decision-making includes analyzing operational data and supporting decision-making and making critical pre-determined decisions to protect the desalination system from physical damage, safeguarding capital investment and improving social impact. Recommendation includes providing recommendations to designers for design improvements, providing recommendations to manufacturers for product improvements, and providing recommendations to economists on balanced water financing and pricing strategies (e.g., the pricing model presented at the holding company exhibition). As shown in Figure 70.

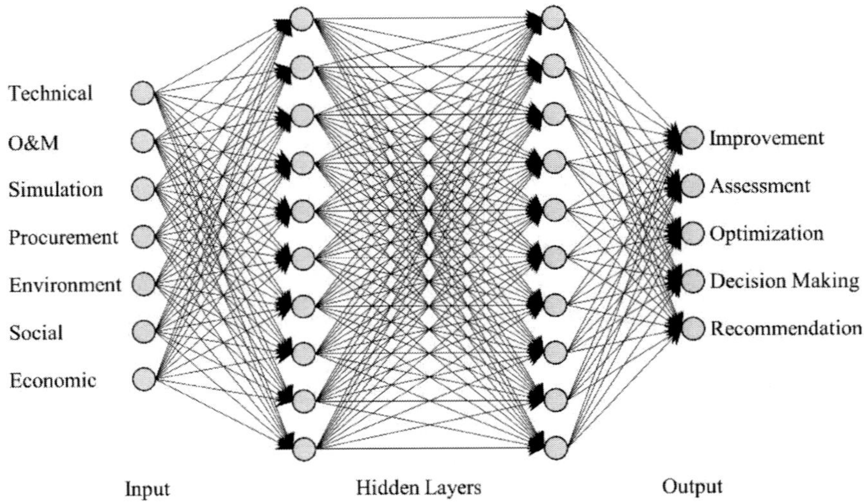

Figure 70. Architecture of AI-Sim Total Cost Reduction and Economics Tool Model.

The cost and economics of desalination aren't just limited to the desalination process itself; they also include the water distribution system. An AI-powered desalination system needs to be thought about to cut down on water loss and energy use by finding leaks, predicting maintenance needs, adjusting flow rates, and managing pressure.

It is possible to lower the cost of making desalination systems by using AI-Sim to predict and forecast the performance of desalination plants using the desalination performance index, model and simulate process bottlenecks, water quality, and plant performance, and improve operation planning and preventative maintenance.

Total Cost Reduction is critical to ensure the economic viability of desalination systems. The AI-Sim total cost reduction and economics tool can optimize desalination system operation to reduce total costs and improve economics.

AI-Sim total cost reduction and economics tool consists of data collection, data preprocessing, a machine learning model, a simulation engine, an optimization algorithm, and economic analysis. Data collection from desalination system operation, including capital cost, operation cost, maintenance costs, consumables costs, currency exchange rates, energy consumption, water production, and historical data; preprocess data to remove noise, handle missing values, normalize data, and feature-engineer data; create

AI/ML models that can predict total costs based on operational parameters and preprocessed data, such as decision trees and neural networks for training; create simulation engines for modeling the desalination process using simulation software; create optimization algorithms to improve the operation of the desalination system to lower total costs; and do economic analysis to figure out costs, ROI, and payback period.

AI-Sim total cost reduction and economics tool strategy is based on optimizing water production, energy consumption, operation and maintenance costs, and the use of everyday items. We optimize energy consumption through methods such as energy demand prediction, pump operation optimization, and membrane efficiency improvement. Similarly, we optimize water production through methods such as water demand prediction, membrane operation optimization, pre-treatment efficiency enhancement, and post-treatment efficiency improvement. Costs of operation and maintenance can be lowered by predicting when equipment will break down, making maintenance schedules more efficient, and making equipment more reliable. Consumables can be used more efficiently by predicting when they will be needed, replacing membranes more efficiently, using spare parts more efficiently, dosing chemicals more efficiently, and so on.

Predictive modeling, anomaly detection, regression analysis, classification, clustering, and deep learning are some of the AI/ML algorithms used in AI-Sim to lower costs and help with economics [238, 239]. The algorithms use predictive modeling to forecast energy consumption and maintenance costs; anomaly detection to identify abnormal operational patterns; regression analysis to model relationships between variables; classification to categorize operational status, such as optimal and suboptimal; clustering and grouping similar operational patterns; and deep learning based on CNN and RNN.

Some of the AI-Sim total cost reduction and economics tools are simulation modules that model the process, membrane fouling, maintenance, and the economy. These modules model the process for things like mass balance, momentum balance, transport phenomena, and more. They also model and simulate to predict membrane fouling, figure out and optimize energy use, use predictive maintenance scheduling, and calculate costs and revenue [240-244].

AI-Sim's total cost reduction and economics tool strategy objected to minimizing total cost, maximizing plant availability, optimizing energy consumption, minimizing maintenance costs, and maximizing recovery. Minimize total cost by reducing energy consumption and maintenance costs;

maximize plant availability through optimizing uptime and productivity; optimize energy consumption by reducing specific energy consumption; minimize maintenance costs through predictive maintenance scheduling; and maximize recovery to optimize water production.

Cost-Benefit Analysis [245], Capital Expenditure (CapEx), Operating Expenditure (OpEx), ROI [246], Payback Period [247], NPV [248], and Levelized Cost of Water (LCOW) [249] are some of the economic analysis modules that can be used for AI-SIM total cost reduction and economies tools. The cost-benefit analysis evaluates the economic viability of the desalination system, CapEx accounts for initial investment costs, operating expenditure (OpEx) includes energy, maintenance, and labor costs, ROI calculates the payback period for the desalination system, NPV assesses the profitability of the desalination project, and LCOW calculates water production costs.

AI-Sim total cost reduction and economics provide a strong way to boost economics by raising ROI, lowering payback period, raising NPV, and lowering total costs. It also improves system availability, system efficiency, energy efficiency, membrane lifetime, recovery, and helps people make better decisions. KPIs measure how well water quality is improved by looking at things like total cost reduction, plant availability, performance index, environmental performance index, energy consumption, maintenance, recovery, specific energy consumption, and levelized cost of water. They also try to keep the desalination system's negative effects on the environment to a minimum.

5. Operation Staff Skills Enhancement

Using AI-Sim to enhance operation staff skills, desalination sustainability is achieved through many frameworks, and operation and maintenance team qualifications are considered as a direct framework that affects desalination. AI-Sim power desalination models can be used to enhance desalination staff skills through coding operation staff members and linking these codes to AI-Sim algorithms, which allow monitoring and evaluating each member's performance, considering members' technical, physical, mental, and health status linked by process performance. Analyzing the data using these algorithms leads to identifying the strengths and weaknesses of each individual member. AI-Sim algorithms are expected to propose a specific training course for each member, as well as propose incentives and rewards related to each member's specific performance as per analysis results, considering operator-recorded enhancement proposals during the work period to enhance staff skills.

AI-Sim operation staff skills enhancement tool can enhance operation staff skills by providing a realistic and interactive learning environment. It can improve staff operation skills in areas such as plant operation and control, troubleshooting and maintenance, energy management and optimization, water quality management, and decision-making. The AI-Sim Operation Staff Skills Enhancement tool has an input and output as follows:

Tool inputs, AI-Sim operation staff skills enhancement tool data acquisition can be collected from six main sources: technical data, O&M data, simulation data, social data, economic data, and legal data.

Technical data includes technical standards for production costs, organizational structures for operating and maintenance teams, and specialized training courses available. O&M data includes; real-time operational data of the desalination system under research and development, historical operation and maintenance data of the desalination system under research and development, real-time operational data of other desalination systems, historical operation and maintenance data of other desalination systems, and initial operation and maintenance plans. Simulation data includes; data generated from simulation and optimization processes. Social data includes; social data of the plant's operational area, operator data, including specialization, previous experience, educational background, and training level, and employees' health and psychological data. Economic data includes; economic data for the desalination system operational area, economic and social level data and wages in the desalination system operational area, and economic standards for production costs. Legal data includes; local labor laws, employee performance evaluation standards, and work regulations for operating and maintenance companies.

Tool outputs, AI-Sim operation staff skills enhancement tool in demand outputs are prediction, optimization, classification, and recommendation. Prediction includes; strengths and weaknesses, training needs and training topics for each employee individually, and future job development expectations for each employee based on predefined measurement criteria. Optimization includes; psychological condition of employees and its impact on operational efficiency, maintenance, and the economics of the desalination plant, and the health condition of employees and its impact on operational efficiency, maintenance, and the economics of the desalination plant. Classification includes; determining the technical efficiency of employees, correlated with technical and economic performance. Recommendation includes; proposals for incentives and rewards, defining the required qualifications for each specialization and future qualification requirements, providing human development proposals to improve employee efficiency and required qualifications, and providing proposals for managers to improve salaries and link them to efficiency based on technical and economic standards. As shown in Figure 71.

AI-Sim operation staff skills enhancement tool consists of data collection, data preprocessing, machine learning model, and simulation environment. The tool collects data from the desalination system in operation, including instrumentation data such as temperature, pressure, and flow rate; equipment data such as pump performance and subsystem performance; operator actions and decisions; and historical data from the system. Data preprocessing involves removing noise, handling missing values, and normalizing the data. Additionally, it involves developing a machine learning model to simulate the operation of the desalination plant and predict outcomes based on operator actions and decisions. Finally, it involves creating a simulation environment that replicates the desalination plant operation, complete with a graphical user interface (GUI), real-time data visualization, and interactive controls.

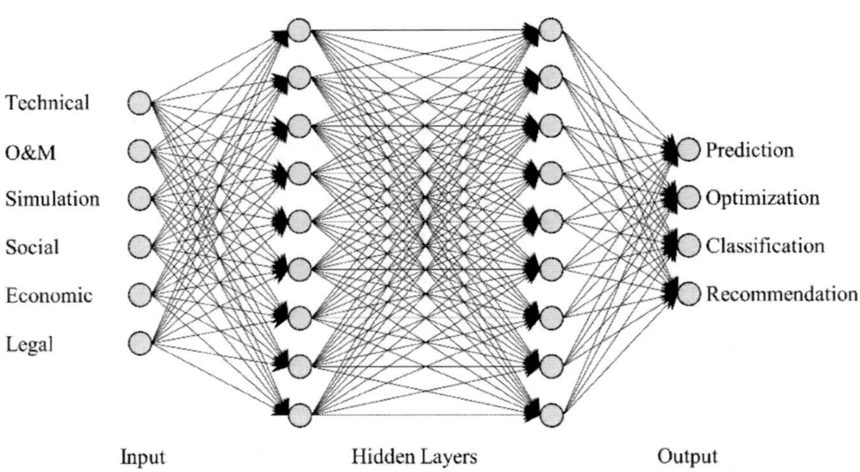

Figure 71. Architecture of AI-Sim Operation Staff Skills Enhancement Tool Model.

GUI is A user-friendly interface that mimics the desalination system in the control room; Real-time data visualization of plant operation, including sensor data, equipment performance, and water quality; Operators can manipulate plant operation through interactive controls, which include starting and stopping equipment, adjusting operating parameters, and responding to alarms [250]. Simulation scenarios such as normal operation scenarios, emergency situation scenarios, and maintenance and repair scenarios have to be considered for operation staff skills enhancement. The normal operation scenario simulates normal plant operations, including start-up, steady-state, and shutdown; the emergency situations scenario replicates emergency situations such as equipment failures, power outages, and water quality issues; and the maintenance and repair scenario replicates routine maintenance, equipment replacement, and troubleshooting. Skills enhancement modules consist of plant operation and control, troubleshooting and maintenance, energy management and optimization, and water quality management.

Plant operation and control to enhance skills in operating and controlling the desalination plant, including starting and stopping equipment, adjusting operating parameters, and responding to alarms and alerts; troubleshooting and maintenance to improve skills in troubleshooting and maintaining equipment, including identifying faults and failures, performing routine maintenance tasks, and troubleshooting complex issues; energy management and optimization to enhance skills in managing and optimizing energy consumption, including monitoring energy usage, identifying energy-saving opportunities, and optimizing energy consumption using advanced technologies; water quality management to improve skills in managing water quality, including monitoring water quality parameters, identifying water quality issues, and implementing corrective actions to maintain water quality.

AI-Sim operation skills enhancement tool encompasses machine learning algorithms, real-time optimization, and what-if analysis. Machine learning algorithms, such as predictive models, classification models, and regression models, are used to simulate plant operation. Real-time optimization improves plant operations, such as energy use, water quality, and equipment performance. And what-if analysis looks at the effects of different operating scenarios, such as changing operating parameters, equipment failures, and water quality problems.

AI/ML algorithms for the AI-Sim operation skills enhancement tool encompass neural networks, decision trees, deep learning, and reinforcement learning. Neural networks are used for fault detection and predictive modeling. Decision trees are used for troubleshooting and making decisions. Clustering is used for pattern recognition and finding outliers. Deep learning is used for modeling and optimizing complex processes. Reinforcement learning is used for making the best decisions and controlling them. Simulation-Based Training for AI-Sim operation skills enhancement tools include virtual plant tours, scenario-based training, fault detection and troubleshooting, optimization and energy efficiency, and emergency response training.

Virtual plant tours include interactive and immersive plant overviews; scenario-based training for realistic operational scenarios and decision-making; fault detection and troubleshooting for identifying, diagnosing, and resolving issues; optimization and energy efficiency strategies for energy reduction; and emergency response training by simulation of emergency situations. AI-Sim operation staff skills enhancement would offer a powerful tool to improve operator skills and knowledge, leading to improved plant operation and efficiency, reduced training time and costs by providing a realistic and interactive learning environment, increased system efficiency by optimizing energy consumption, water quality, and equipment performance, enhanced safety by training operators to respond to emergency situations and alarms, reduced energy consumption, enhanced troubleshooting skills, increased plant availability, better decision-making, reduced risk of non-compliance with regulations, and cost savings by reduced energy and maintenance costs.

KPIs are used to evaluate operational staff skills enhancement, such as plant availability, energy consumption, membrane lifetime, troubleshooting time, decision-making accuracy, compliance with regulations, system performance index, and environmental performance index.

6. Enhancing Desalination System Design through Practical Applications

The practical enhancement of desalination systems has different areas, such as process optimization, membrane fouling mitigation, energy recovery, plant design, and operational troubleshooting. Process optimization improves permeate quality and reduces energy consumption; membrane fouling mitigation extends membrane lifespan and reduces cleaning frequency; energy recovery optimizes energy consumption and reduces waste; plant design optimizes layout and reduces capital costs; and operational troubleshooting expedites issue identification and resolution.

AI-Sim Practical Enhancement tool can enhance desalination system operation, maintenance, and design. It can provide practical enhancements in desalination system efficiency, productivity, and sustainability. AI-Sim practical enhancement tool has an input and output as follows:

Tool inputs, AI-Sim practical enhancement tool data acquisition can be collected from five main sources: technical data, O&M data, simulation data, environmental data, and economic data. Technical data includes design data for the desalination plant, technical data for the components used in the desalination system, historical energy consumption data, and data on the expected lifespan of equipment. O&M data includes real-time operational data of the desalination system under research and development, historical operation and maintenance data of the desalination system under research and development, real-time operational data of other desalination systems, historical operation and maintenance data of other desalination systems, initial operation and maintenance plans, and record of faults and how they were repaired. Simulation data includes data generated from simulation and optimization processes. Environmental data includes waste data and allowable environmental standards. Economic data includes historical economic cost data.

Tool outputs, AI-Sim practical enhancement tool in demand outputs are prediction, improvement, optimization, and recommendation.

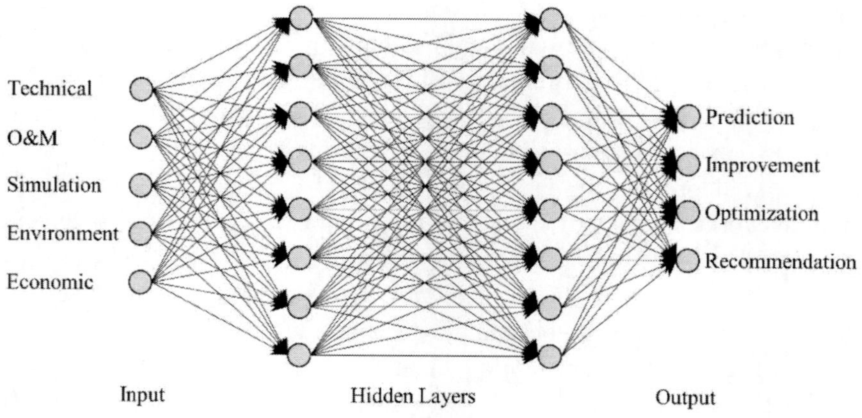

Figure 72. Architecture of AI-Sim Practical Enhancement Tool Mode.

Prediction includes predicting and identifying operational faults and suggestions for their repair. Improvement includes; improving waste quality, and Extending equipment lifespan. Optimization includes proposals to improve desalination processes based on existing designs and actual operational and maintenance practices, improving energy consumption based on the actual desalination system performance, and reducing economic costs. Recommendation includes providing suggestions to designers to improve designs according to practical realities, and providing suggestions to manufacturers to improve products based on actual operational results. As shown in Figure 72.

AI-Sim practical enhancement tool consists of data collection, data preprocessing, machine learning model, and simulation environment. Collect data from desalination system operation, including instrumentation measurement data, equipment performance, and water quality; preprocess data to get rid of noise, deal with missing values, and normalize data; create a machine learning model to simulate how a desalination system works and guess what will happen. Create a simulation environment that functions similarly to a desalination system, allowing individuals to conduct tests and experiments. Numerous areas, such as energy efficiency, water quality, equipment maintenance, system design, and operator training, offer practical enhancements. Energy efficiency: to predict energy demand, improve pump operation, and make membranes more efficient; improving water quality: to predict water quality parameters, improve treatment processes, and find anomalies and faults; and equipment maintenance: to predict equipment failures and make the best maintenance schedules by looking at equipment performance data, analyzing it, finding anomalies and faults, and planning maintenance. system design optimization for improved efficiency, productivity, and sustainability by different design scenarios simulation, design options evaluation, and design parameters optimization; operator training to enhance skills using simulated scenarios.

AI/ML algorithms for the AI-Sim practical enhancement tool include neural networks, decision trees, clustering, deep learning, and reinforcement learning. There are different types of neural networks, each with its own purpose. Predictive modeling and fault detection are two examples. Decision trees and clustering are two examples of how to fix problems and make decisions. Deep learning and reinforcement learning are two examples of how to model and improve complex processes. The Simulation Modules for the AI-Sim Practical Enhancement tool include process simulation, membrane fouling simulation, energy consumption simulation, plant layout simulation, and fault detection simulation.

The AI-Sim Practical Enhancement tool uses process simulation to dynamically model the desalination process, membrane fouling simulation to model fouling mechanisms, energy consumption simulation for calculation and optimization, plant layout simulation for plant design optimization, and fault detection simulation for troubleshooting and diagnosis. Simulation Techniques for AI-Sim, a practical enhancement tool, encompass predictive modeling, optimization algorithms, machine learning, and simulation-based optimization. The tool utilizes predictive modeling to forecast the operation of the desalination system and predict its outcomes; optimization algorithms to enhance its efficiency; machine learning techniques to analyze data and make predictions; and simulation-based optimization to further optimize the system. AI-Sim practical enhancement would provide a promising powerful tool that can improve desalination system efficiency and reduce costs, enhance desalination system productivity and improve water quality, increase desalination system sustainability and reduce environmental impact, provide insights and support better decision-making, increase system availability, reduce energy consumption, increase membrane lifetime, improve permeate quality, and reduce capital and maintenance costs.

KPIs are used to evaluate practical enhancements such as energy consumption, membrane lifetime, permeate quality, system availability, maintenance cost, recovery, system performance index, and environmental performance index. AI-Sim practical enhancement can be used to improve the site selection process for new desalination projects by looking at a lot of data, such as geographical, environmental, and socio-economic factors, and correctly predicting the best places for these projects to be built.

In the context of technology selection, thermal and membrane have distinct advantages and limitations. Evaluating and choosing a suitable process becomes essential. Technology selection based on specific characteristics is a critical step in desalination project planning and development. AI-Sim's practical enhancement application using deep analysis will help streamline the process.

In the context of operation, the desalination process continuously monitors in real time for various parameters using sensors and Internet of Things (IoT), supporting optimal conditions throughout the desalination process. AI-Sim, a practical enhancement, powers the desalination process with practical algorithms that analyze real-time data to ensure membrane integrity and monitor water quality.

7. Characteristic Equations Coefficients Enhancement

Using AI-SIM to enhance equipment manufacturer characteristic constants in equations as per real field operating conditions. Equipment characteristic equations explain the relationship governing equipment input and output variables. characteristic equations contain some coefficients that are used to express the equipment performance as per design parameters and operating conditions. These coefficient values are determined by experiments and laboratory measurements to create what we call an empirical equation. For instance, the performance equation that governs centrifugal pumps illustrates the connection between pump flow and pressure during pump operation. The relationship between pump flow and pressure is shown in Equation (74) by the pump pressure (head) being equal to a coefficient (k_1), the pump flow is multiplied by a coefficient (k_2), and the pump flow is multiplied by a coefficient (k_3).

$$H = k_1 - k_2 Q - k_3 Q^2 \tag{134}$$

where: H: Head, (m); Q: Flowrate, (m^3/s); and k_1, k_2, k_3 are coefficients of the pump, impeller geometry and operation speed (k_1 (m), k_2 (s/m^2), and k_3 (s^2/m^5).

These coefficient values depend on pump impeller geometry, operation speed, internal part surface roughness, spacing between internal mechanical parts, and other elements. Initial coefficient values introduced by the pump manufacturer express the pump characteristic at a specific fluid type, specific speed, new mechanical components, etc., such that the pump runs into operation, considering the pumped fluid's real characteristics and operating pump speed, which may be changed from time to time, in addition to natural wear in the internal mechanical parts as a result of long operation time. The pump coefficient values change from the initial coefficient values introduced by the pump manufacturer, which means that the manufacturer-issued performance curves become not representative of the actual pump performance. When AI is combined with simulation (AI-Sim), real-time centrifugal pump characteristics can be given by coefficients reverse determination results. This is done by measuring and recording real-time flow and pressure operating data in different cases using AI. Then, simulation integration is used to finish drawing the pump characteristic curve.

AI-Sim tools can be made to guess the best conditions for operation, plan preventative maintenance ahead of time, warn of the downsides of operating in certain conditions, warn of mechanical failure ahead of time, and guess the economic effects of each situation. Meanwhile, feedback designers and researchers gain research proposals or design improvements from their experiences and simulation results. Desalination systems rely on complex equations to model their behavior. Characteristic equations describe the relationships between system variables, such as pressure, flow rate, salinity, and temperature. Enhancing coefficients in characteristic equations is crucial to ensuring accurate modeling and simulation to improve operation, performance, and sustainability. Desalination systems contain many equipment, each equipment has its own characteristic equations based on its individual behavior which affect the system behavior, these equations such as flow equations and pressure drop equations that describe the pressure drop across various components of the desalination system, such as pipes, valves, and membranes, mass transfer equations that describe the transport of solutes across the membrane, including rejection and permeability, heat transfer equations that describe the transfer of heat across various components of the desalination system, including pipes, valves, and membranes, momentum balance equation that models fluid dynamics and pressure drop, transport phenomena equation that describes water and solute transport and membrane fouling equation that models fouling mechanisms.

Characteristic equation coefficient enhancement is essential to accurately model, simulate, and enhance desalination system operation, such as mass transfer coefficient, which affects solute rejection and flux; membrane permeability, which influences water flux and salt rejection; fouling resistance, which impacts membrane lifespan and cleaning frequency; hydraulic resistance, which affects pressure drop and energy consumption; and solute rejection coefficient, which influences permeate quality. The AI-Sim characteristic equation coefficients enhancement framework consists of data collection, data preprocessing, machine learning model, and simulation environment.

Tool inputs, AI-Sim characteristic equations coefficients updating tool data acquisition can be collected from five main sources: technical data, O&M data, simulation data, procurement data, and economic data. Technical data includes governing equations for all desalination system components, technical specifications for the desalination system components, performance efficiency standards for the desalination system components, raw water specifications, and designer data. O&M data includes real-time operational data of the desalination system under research and development, historical operation and maintenance data of the desalination system under research and development, real-time operational data of other desalination systems, historical operation and maintenance data of other desalination systems, and initial operation and maintenance plans. Simulation data includes data generated from simulation and optimization processes. Procurement data includes manufacturer data for the equipment. Economic data includes historical economic cost data.

Tool outputs, AI-Sim characteristic equations coefficients updating tool in demand outputs are improvement, optimization, and recommendation.

Improvement includes, developing maintenance plans according to updated equation parameters based on actual operational data, and improving operational and maintenance plans according to updated parameters. Optimization includes, updating parameters according to the actual operating conditions, and Improving equipment equation parameters. Recommendation includes, providing designers with suggestions for design improvements, providing manufacturers with suggestions for improving equipment manufacturing, providing suggestions to designers to improve design, and Providing suggestions to manufacturers to improve products. As shown in Figure 73.

The framework collects data from the desalination system operation, which includes instrumentation measurements and system performance metrics; preprocesses the data to remove noise, handle missing values, and normalize the data; develops a machine learning model to predict coefficients in characteristic equations based on system operating conditions; and creates a simulation environment to integrate the updated characteristic equations and coefficients.

AI/ML algorithms for enhancing the coefficients of the AI-Sim characteristic equations include neural networks, genetic algorithms, particle swarm optimization, deep learning, and reinforcement learning. Neural networks are utilized for predictive modeling and coefficient estimation, genetic algorithms for coefficient optimization, particle swarm optimization for hyperparameter tuning, deep learning for complex process modeling, and reinforcement learning for optimal control and coefficient enhancement. Simulation modules for the AI-Sim characteristic equations coefficient enhancement tool include process simulation, membrane fouling simulation, energy consumption simulation, plant layout simulation, and a coefficient enhancement module. Process simulation for desalination process dynamic modeling; membrane fouling simulation for fouling mechanisms modeling; energy consumption simulation such as calculation and optimization; plant layout simulation for plant design optimization; coefficient enhancement module for coefficient estimation. Simulation techniques include neural networks, genetic algorithms, and particle swarm optimization. Neural networks predict coefficients in characteristic equations based on system operating conditions, genetic algorithms optimize coefficients in characteristic equations for improved accuracy and robustness, and particle swarm optimization optimizes coefficients in characteristic equations for improved accuracy and convergence.

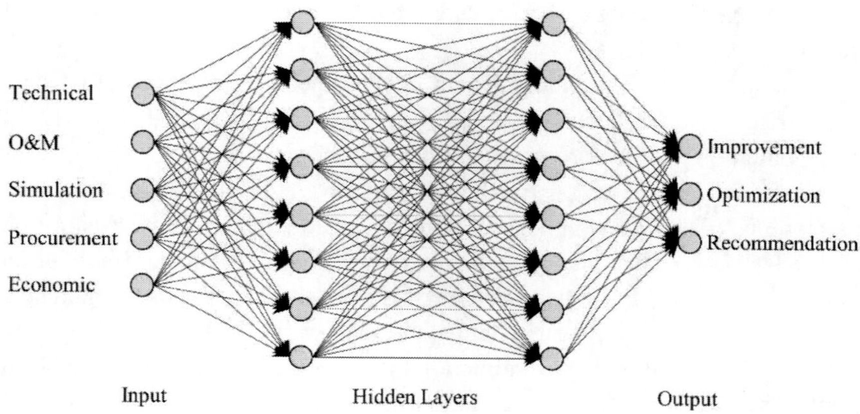

Figure 73. Architecture of AI-Sim Characteristic Equations Coefficients Updating Tool Model.

It is possible to improve reliability by relying less on empirical correlations and having more faith in simulation results. It is also possible to improve efficiency by cutting down on computational time and effort, improve accuracy by updating coefficients to improve simulation reliability, make better decisions by using data-driven insights for optimization, improve plant efficiency by optimizing coefficients that lower energy consumption, extend membrane lifetime by accurate fouling modeling, and improve scalability by using updated coefficients for new plant designs.

KPIs are used to describe improvements in equation coefficients. Some examples are accuracy of the coefficients, simulation reliability, system efficiency, membrane lifetime, energy use, permeate quality, system performance index, and environmental performance index.

8. Research and Development Enhancement

In the era of artificial intelligence, we must develop two main frameworks for AI-Sim, research and development in the desalination industry. The first framework is to enhance the desalination industry technically and economically using AI in integration with Simulation AI-Sim. The second framework is to develop AI algorithms to enhance AI-Sim can be used in desalination industry research and development (R&D), i.e., Using AI-Sim for desalination R&D AI-Sim techniques. Desalination systems require continuous R&D to improve efficiency, reduce costs, and enhance water quality. AI-Sim Research and Development enhancement tool has an input and output as follows:

Tool inputs, AI-Sim research and development enhancement tool data acquisition can be collected from five main sources: technical data, O&M data, simulation data, environmental data, and economic data.

Technical data includes technical data for the desalination system components, data on technical obstacles, and technical standards to be achieved. O&M data includes real-time operational data of the desalination system under research and development, historical operation and maintenance data of the desalination system under research and development, real-time operational data of other desalination systems, historical operation and maintenance data of other desalination systems, and initial operation and maintenance plans. Simulation data includes data generated from simulation and optimization processes. Environmental data includes environmental standards to be achieved. Economic data includes data on economic obstacles, economic standards to be achieved, and technical and economic feasibility studies.

Tool outputs, AI-Sim research and development enhancement tool in demand outputs are improvement, optimization, and recommendation. Improvement includes improving research and development outcomes for desalination system components, and proposals for developing artificial intelligence in integration with simulation. Optimization includes developing research plans within a technical and economic framework, developing plant management plans, and developing economic and financial management. Recommendation includes data on technical topics requiring research, list of required technical and economic studies, providing suggestions to designers for design improvement, and providing suggestions to manufacturers for product improvement. As shown in Figure 74.

AI-Sim research and development enhancement tool can accelerate R&D in desalination systems by reducing the need for physical prototyping and experimentation. The AI-Sim research and development enhancement framework should consist of data collection, data preprocessing, machine learning models, and simulation environments. In this framework, data is collected from the desalination system's operation, including instrumentation measurements and system performance metrics. The data is then preprocessed to get rid of noise, deal with missing values, and normalize it. A machine learning model is then built to simulate the system's behavior and predict what will happen. Finally, a simulation environment is made to integrate the machine learning model and make testing and experiments easier.

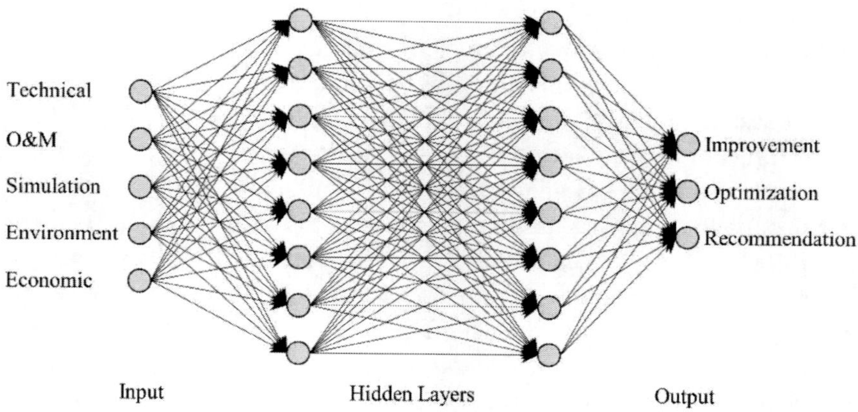

Figure 74. Architecture of AI-Sim Research and Development Enhancement Tool Model.

AI-Sim research and development enhancement tools can be used for desalination R&D enhancement in areas such as membrane development, system design optimization, energy recovery systems, and water quality improvement. Membrane development to design and optimize new membrane materials and structure's durability, performance improvement and sustainability; system design optimization to optimize desalination system design for efficiency, productivity improvement, and sustainability for RO, MSF, MED, …etc.; energy recovery for energy recovery systems optimization and reduce energy consumption; water quality improvement for treatment processes optimization and improve water quality, include optimization of coagulation, flocculation, sedimentation, filtration, disinfection and remineralization, which leads to membrane innovation to improve material properties, structures; process optimization by enhanced recovery rates, reduced energy consumption; fouling mitigation to use advanced cleaning methods, fouling-resistant membranes; scaling prevention using predictive modeling, and anti-scaling technologies; energy consumption optimization, and create novel energy-recovery systems. AI/ML algorithms for AI-Sim research and development enhancement tools include predictive modeling, optimization algorithms, reinforcement learning, and data analytics.

Predictive modeling for performance forecasting and parameter optimization; optimization algorithms such as genetic, particle swarm, and simulated annealing; deep learning through neural networks and CNN; reinforcement learning in the context of autonomous optimization and decision-making; and data analytics for pattern recognition and anomaly detection. Simulation modules for AI-Sim research and development enhancement tools should include simulation of process, membrane fouling, energy consumption, plant layout, scaling, and fouling.

The process simulation can be used to dynamically model the desalination process; the membrane fouling simulation is used to model the fouling mechanisms; the energy consumption simulation is used for calculation and optimization; the plant layout simulation is used for plant design optimization; and the scaling and fouling simulation is used for predictive modeling. Simulation techniques include neural networks, genetic algorithms, and particle swarm optimization. Neural networks, like feedforward neural networks, recurrent neural networks, and convolutional neural networks, model how desalination systems work and predict what will happen. Genetic algorithms improve the design and operation of desalination systems by doing things like population initialization, fitness evaluation, selection and crossover, and mutation and elitism. Particle swarm optimization improves the operation of desalination systems and lowers energy use by doing things like particle initialization, velocity update, position update, and fitness evaluation.

AI-Sim research and development enhancement is a powerful tool that speeds up research and development in desalination systems by reducing the need for physical prototypes and experiments. This leads to shorter development cycles and better desalination system efficiency by optimizing design and operation. It also lowers energy use, improves water quality by optimizing treatment processes and predicting water quality parameters, and helps people make decisions by giving them data-driven insights. KPIs include membrane permeability, water recovery rate, product water quality, fouling resistance, scaling propensity, plant availability, specific energy, system performance index, and environmental performance index. They are used to study and make improvements in these areas.

9. Process Design Optimization

Desalination systems require optimal process design to achieve efficient and sustainable operation. The AI-Sim process design optimization tool can improve the design of desalination systems by looking at how different variables interact with each other. It can also improve process design parameters, like feedwater flow rate, pressure, and temperature, to get the best recovery, the best membrane configuration, the least amount of energy use, the best pump operation, lower capital costs, make the water better, and make the plant more available. AI-Sim process design optimization tool has inputs and outputs as follows:

Tool inputs, AI-Sim process design optimization tool data acquisition can be collected from four main sources: technical data, O&M data, simulation data, and economic data. Technical data includes; technical specifications of desalination system components, efficiency measurement standards for the desalination process, raw water specifications, produced water specifications, technical design standards, data on applied and available desalination processes, and data on research and studies. O&M data includes real-time operational data of the desalination system under research and development, historical operation and maintenance data of the desalination system under research and development, real-time operational data of other desalination systems, historical operation and maintenance data of other desalination systems, and initial operation and maintenance plans. Simulation data includes data generated from simulation and optimization processes. Economic data includes economic standards to be achieved, and technical and economic feasibility studies.

Tool outputs, AI-Sim process design optimization tool in demand outputs should be improvement, optimization, and recommendation.

Improvement includes improving design standards, producing design standard maps for desalination system components based on actual operating conditions, and improving hydraulic balance and stability. Optimization includes enhancing design standards according to real operational data and the technical capabilities of components, determining the optimal technical and economic selection for sub-processes, identifying the best choice for desalination system components and equipment, and providing the best desalination process based on raw water specifications, produced water specifications, desalination system capacity, and location. Recommendation includes; providing research and study centers with required topics for investigation, providing suggestions to designers for design improvement, and providing suggestions to manufacturers for product improvement. As shown in Figure 75.

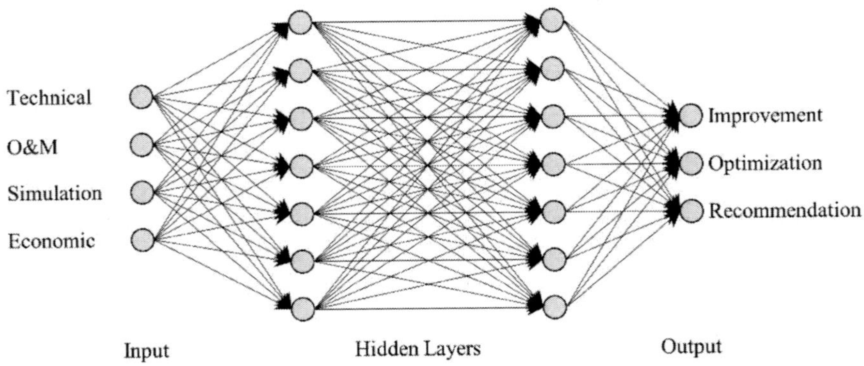

Figure 75. Architecture of AI-Sim Process Design Optimization Tool Model.

AI-Sim optimal process design consists of data collection, data preprocessing, machine learning models, and simulation environments. The steps in the process are: collecting data from the desalination system's operation, such as instrumentation measurements and system performance metrics; preprocessing the data to get rid of noise, deal with missing values, and normalize the data; creating a simulation environment to test and experiment with the machine learning model; and learning how to simulate the behavior of the desalination system.

AI-Sim optimal process design tool can be used for desalination process enhancement in areas such as intake and outfall, pretreatment process, reverse osmosis system design, energy recovery system design, post-treatment process, and water quality improvement. Intake and outfall design optimization including intake tower, piping, intake sump hydraulics and outfall dilution methods; pretreatment processes optimization, include coagulation, flocculation, sedimentation, filtration, …etc.; reverse osmosis system design parameters optimization such as membrane type, module configuration, RO skid hydraulic optimization, operating pressure, and membrane development to design and optimize new membrane materials and structures durability; energy recovery systems design parameters optimization, such as pressure exchanger, hydraulic turbocharger and their efficiencies; post-treatment processes optimization, including disinfection, remineralization, storage and water quality improvement.

AI/ML algorithms for the AI-Sim optimal process design tool includes neural networks, genetic algorithms, particle swarm optimization, deep learning, and reinforcement learning. Genetic algorithms are used for multi-objective optimization, particle swarm optimization through hyperparameter tuning, deep learning is used for complex process modeling, and reinforcement learning is used for autonomous optimization. Neural networks are used for predictive modeling and optimization. The AI-Sim optimal process design tool has simulation modules for coming in and going out, pretreatment, desalination processes, membrane fouling, energy use, post-treatment, and plant layout. An intake and outfall simulation is used for hydraulic and physical modeling. A pretreatment simulation is used to model pretreatment processes. A desalination process simulation is used to model the desalination process dynamically. A membrane fouling simulation is used to model fouling mechanisms. An energy consumption simulation is used to calculate and improve energy use. A posttreatment simulation is used to model posttreatment processes. A plant layout simulation is used to improve plant design.

Simulation techniques include neural networks, genetic algorithms, particle swarm optimization, and deep learning. Neural networks are utilized to model and predict the operation of a desalination system, which includes selecting the intake and outfall configuration, pretreatment process, desalination process, membrane type and configuration, operating pressure and flow rate, pump operation, energy recovery devices, post-treatment process, plant layout, and equipment. Genetic algorithms and particle swarm optimization to improve the efficiency and sustainability of process design parameters. Deep learning methods, like CNNs and RNNs, are used to look at large amounts of data and make desalination systems work better.

AI-Sim optimal process design offers a powerful tool for desalination system efficiency improvement by optimizing process design parameters for desalination system efficiency improvement, design optimization, operation optimization, and energy consumption reduction; sustainability enhancement to optimize process design parameters for sustainability improvement and environmental impact reduction; increase water productivity and improved water quality; enhances water quality by optimizing treatment processes and predicting water quality parameters; optimizes process design parameters for reduced costs by optimizing energy consumption, and improved economic viability; accelerates R&D in desalination systems by reducing the need for physical prototyping and experimentation. KPIs are used to optimize process designs such as capital costs, specific energy consumption, recovery, water quality parameters, plant availability, system performance index, and environmental performance index.

10. Minimize Negative Environmental Impact

Although desalination systems provide freshwater, they can harm the environment if not designed and operated sustainably, so they need to minimize negative environmental impact to achieve efficient and sustainable operation. The AI-Sim minimize negative environmental impact tool can help minimize negative environmental impact by optimizing desalination system design and operation and achieving sustainability goals, such as reducing energy consumption, minimizing waste, and protecting marine life. AI-Sim algorithms can accurately predict the potential ecological consequences of establishing a desalination plant at a given site by analyzing environmental factors. Analysis results using AI-Sim algorithms enable decision makers to choose locations with minimal negative environmental impact. AI-Sim minimize negative environmental impact tool has inputs and outputs as follows:

Tool inputs, AI-Sim minimize negative environmental impact tool data acquisition can be collected from four main sources: technical data, O&M data, simulation data, and environmental data.

Technical data includes; technical specifications of desalination system components. O&M data includes; real-time operational data of the desalination system under research and development, historical operation and maintenance data of the desalination system under research and development, real-time operational data of other desalination systems, historical operation and maintenance data of other desalination systems, and initial operation and maintenance plans. Simulation data includes; data generated from simulation and optimization processes. Environmental data includes; environmental performance measurement standards for the desalination system, environmental description of the area surrounding the desalination system, local and international environmental standards and regulations, and initial environmental impact resulting from construction, operation, and maintenance activities.

Tool outputs, AI-Sim minimize negative environmental impact tool in demand outputs are improvement, optimization, risk prediction, and recommendation. Improvement includes; developing scenarios to mitigate negative impacts according to their priority level, suggesting modifications to existing designs to minimize environmental impact, improving methods for measuring environmental standards, and Developing proposals to raise environmental awareness among desalination system workers and the surrounding community. Optimization includes; proposing optimal operational and maintenance practices to reduce environmental impact. Risk prediction includes; predicting negative impacts on the surrounding environment, and identifying challenges in ranking environmental impacts based on the nature of the surrounding environment at the desalination system location. Recommendation includes; providing implementers, designers, and decision-makers with key considerations for new projects, proposing amendments to environmental laws based on the technologies used and the environmental characteristics of each area, supplying research centers with topics that require solutions or in-depth studies, and providing educational and research institutions with recommendations for developing environmental curricula. As shown in Figure 76.

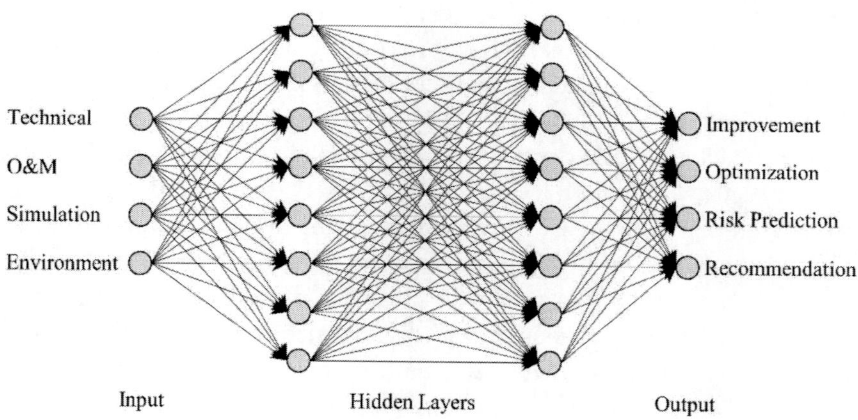

Figure 76. Architecture of AI-Sim Minimizes Negative Environmental Impact Tool Model.

The AI-Sim negative environmental impact framework consists of data collection, data preprocessing, machine learning model, simulation model, environmental impact assessment, and optimization and analysis. Data collection by collecting environmental data, including seawater quality, marine life, and coastal ecosystems; data from desalination system operation, including energy consumption, water production, and waste generation; and external data, such as weather forecasts, ocean currents, and water demand; information preprocessing to remove noise, handle missing values, normalize data, and reduce data dimensionality; machine learning model to simulate desalination system behavior and predict environmental impact by optimizing desalination system design and operation for minimal environmental impact; environmental impact assessment includes life cycle assessment, environmental risk assessment, and developing and tracking sustainability metrics, such as water footprint, energy footprint, and waste reduction; optimization and analysis to optimize desalination system design and operation for minimal environmental impact through performing sensitivity analysis to identify key parameters that affect environmental impact and uncertainty analysis to quantify the uncertainty associated with environmental impact.

AI-Sim minimize negative environmental impact tool can be used to improve the environmental impact of desalination systems in areas like intake and discharge to protect marine life by improving designs like velocity caps and diffusers; energy consumption to lower greenhouse gas emissions and reliance on fossil fuels; cutting down on chemical use for pre-treatment, cleaning, and post-treatment to protect marine life and human health; replacing membranes more efficiently and recycling them; and managing waste more effectively, including getting rid of concentrates and sludge.

AI/ML algorithms in AI-Sim reduce the damage that humans do to the environment. The tool has predictive modeling, optimization algorithms, life cycle assessment, multi-objective optimization, and monitoring based on machine learning. The tool uses predictive modeling to guess how it will affect the environment, optimization algorithms to cut down on chemical and energy use, life cycle assessment to see how the plant will affect the environment over its entire life, multi-objective optimization to balance environmental concerns with operational efficiency, and machine learning-based monitoring to find problems and guess when maintenance is needed. Simulation modules for AI-Sim minimize negative environmental impact; tools include simulation of environmental impact, energy consumption, chemical usage, membrane replacement, and concentrate disposal.

The AI-Sim simulation modules use environmental impact assessment to look at how intake and discharge affect the environment, energy consumption simulation to find the best energy recovery systems, chemical usage simulation to model the need for pretreatment and cleaning chemicals, membrane replacement simulation to guess how long a membrane will last, and concentrate disposal simulation to look at different ways to deal with brine waste. Simulation techniques include Machine Learning Algorithms, Optimization Techniques, and Simulation Software. Machine learning algorithms, like neural networks and decision trees, model and simulate the behavior of desalination systems and predict how they will affect the environment. Optimization techniques, like genetic algorithms and particle swarm optimization, make sure that the design and operation of desalination systems have the least possible effect on the environment.

AI-Sim can minimize negative environmental impact and offers a powerful tool to minimize negative environmental impact by optimizing desalination system design and operation, leading to reduced environmental impact; helps achieve sustainability goals, such as reducing energy consumption and minimizing waste, leading to sustainability improvement; helps to reduce costs and cost saving by optimizing energy consumption, chemical usage, water consumption, and waste generation; and ensures regulatory compliance enhancement by predicting and mitigating environmental harm.

KPIs are used to reduce negative effects on the environment, like harm to marine life during intake and discharge, energy use, chemical use, membrane replacement rate, concentrate disposal volume, carbon footprint, system performance index, and environmental performance index.

11. Social Impact Enhancement

Although desalination systems provide freshwater, they can have social impacts, such as affecting local communities and industries. By optimizing the design and operation of desalination systems, we can achieve Social Impact Enhancement (SIE) goals such as improving public health, supporting local economies, and promoting social justice. Desalination systems have social impact areas such as water accessibility, job creation, community engagement, public health, and environmental awareness.

Water accessibility ensures equitable distribution and affordability; job creation and local employment opportunities; community engagement through stakeholder involvement and education; public health through safe drinking water and reduced water-borne diseases; and environmental awareness through education on sustainable water management. Social impact assessment framework consists of stakeholder identification, social impact metrics, and risk assessment. Identify stakeholders, including local communities, industries, and governments; develop and track social impact metrics, such as public health outcomes, economic benefits, and social justice indicators; and assess risks associated with desalination system operation, including environmental and social risks. AI-Sim Social Impact Enhancement tool has inputs and outputs as follows:

Tool inputs, AI-Sim social impact enhancement tool data acquisition can be collected from six main sources: technical data, O&M data, simulation data, procurement data, social data, and economic data.

Technical data includes, data on water culture and water consumption behavior of the surrounding community, and data on pure water sources relied upon by the surrounding community. O&M data include; real-time operational data of the desalination system under research and development, historical operation and maintenance data of the desalination system under research and development, real-time operational data of other desalination systems, historical operation and maintenance data of other desalination systems, and initial operation and maintenance plans. Simulation data includes data generated from simulation and optimization processes. Procurement data includes data on available raw materials and natural resources in the plant's location. Social data includes social data of the area where the plant is to be established and surrounding regions, data on educational levels and technical specializations of the surrounding community, and data on available professions in the area of the plant's establishment. Economic data includes economic data of the area where the plant is to be established and surrounding regions, data on wages in the surrounding area, and data on water tariff rates in the region.

Tool outputs, AI-Sim social impact enhancement tool in demand outputs are improvement, optimization, and recommendation. Improvement includes developing awareness programs about the nature and social benefits of desalination plants based on the social conditions of the plant's operational area, providing plans to raise social awareness according to the social conditions of the plant's operational area, and presenting plans to clarify the quality of desalinated water and its compliance with local and international health standards, explaining why it sometimes differs from natural water. Optimization includes providing data on locally available materials and resources that can create jobs and employment opportunities, informing designers and implementers about social requirements that may impact design and execution to ensure stability in the desalination sector, providing suggestions to designers for developing designs that align with the surrounding community, and offering proposals to implementers on utilizing local raw materials and resources. Recommendation includes; proposing industries and commercial activities necessary for the establishment, operation, and maintenance of desalination plants to improve the economic level of the plant's operational area, developing training programs to enable the local community to secure jobs in the desalination sector, enhancing technical and economic levels, and suggesting ways to reduce social concerns about the high cost of desalinated water production, along with recommendations for water consumption efficiency. As shown in Figure 77.

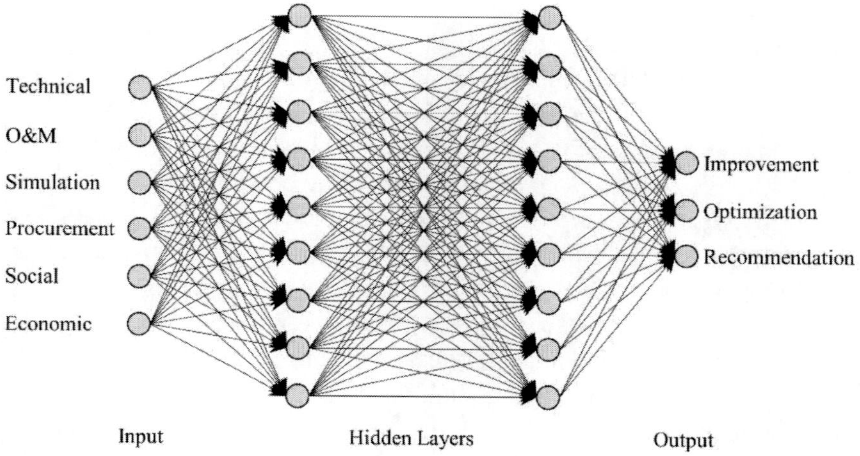

Figure 77. Architecture of AI-Sim Social Impact Enhancement Tool Model.

AI-Sim social impact enhancement consists of data collection, data preprocessing, machine learning model, simulation model, social impact assessment, optimization, and enhancement.

Data collection for stakeholders, including local communities, industries, and governments, social impacts including public health, economic benefits, and social justice indicators, and desalination system operation, including water quality, energy consumption, and waste generation; data preprocessing include data cleaning to clean and preprocess data to remove noise, handle missing values, and normalize data, data transformation to transform data into suitable formats for AI-Sim, and data reduction to reduce data dimensionality using techniques; machine learning model to simulate desalination system behavior and predict social impacts using optimization techniques, such as genetic algorithm and particle swarm optimization, to optimize desalination system design and operation for social impact enhancement, using simulation software to model and simulate desalination system behavior and social impacts; social impact metrics to develop and track social impact metrics, such as public health outcomes, economic benefits, and social justice indicators, stakeholder engagement to identify social impact concerns and priorities, and risk assessment associated with desalination system operation, including environmental and social risks; optimization techniques to be used to optimize desalination system design and operation for social impact enhancement; enhancement by identify opportunities to enhance social impact, including improving public health, supporting local economies, and promoting social justice.

AI-Sim social impact enhancement tool can be used to enhance public health, local economies, and social justice. By designing and operating desalination systems in the best way possible, we can improve public health by lowering the number of waterborne diseases and making the quality of the water better. We can also help local economies by doing things like creating jobs, boosting economic growth, and supporting local industries. Finally, we can promote social justice by making sure that everyone has equal access to water, protecting vulnerable groups, and upholding human rights.

AI-Sim social impact enhancement tool uses machine learning and predictive modeling to do things like social network analysis, multi-objective optimization, predictive modeling, and machine learning-based monitoring. Predictive modeling for Forecast water demand to optimize distribution; Optimization algorithms to Maximize efficiency, minimize costs; Social network analysis to identify community influencers and optimize engagement; and machine-learning-based monitoring to detect anomalies and predict maintenance. minimize costs; social network analysis to identify community influencers and optimize engagement; and machine-learning-based monitoring to detect deviations and predict maintenance. Multi-objective optimization to Balance social, environmental, and economic concerns. Simulation modules for the AI-Sim social impact enhancement tool include water distribution simulation, demand forecasting, community engagement simulation, public health simulation, and environmental awareness simulation.

AI-Sim social impact enhancement tool uses water distribution simulation to improve pipeline networks, demand forecasting to guess how people will use water, community engagement simulation to show how stakeholders will interact, public health simulation to check the effects on water quality, and environmental awareness simulation to check the effectiveness of education programs. AI-Sim social impact enhancement offers a powerful tool to enhance social impact by optimizing desalination system design and operation, improving public health by reducing waterborne diseases and improving water quality, supporting local economies by creating jobs, stimulating economic growth, and promoting local industries, and promote social justice by ensuring equitable access to water, protecting vulnerable populations, and promoting human rights. Optimization Parameters such as water distribution network design, pricing strategies, community engagement initiatives, public health education programs, environmental awareness campaigns, local employment opportunities, and stakeholder involvement are needed to enhance desalination's social impact.

KPIs are used to enhance social impact, such as water accessibility as per population, job creation, community engagement, public health, environmental awareness, water affordability, the social return on investment, the system performance index, and the environmental performance index. The incorporation of AI-powered chatbots and virtual assistants is enhancing customer service and enabling more effective demand management by water utilities. Ultimately, the integration of AI in desalination is not merely a technological breakthrough but represents a vital step towards addressing one of the most pressing challenges of our time: ensuring a sustainable water supply for future generations.

12. Information and Operational Technologies Convergence

Desalination systems require convergence of Information Technology (IT) and Operational Technology (OT) to optimize operation and maintenance, economics, design, and environmental and social impact. AI-Sim IT/OT convergence can facilitate IT/OT convergence by integrating data from various sources and simulating desalination system behavior, which enables real-time monitoring, predictive maintenance, and optimized operation, leading to improved efficiency, reduced costs, and enhanced decision-making. AI-Sim Information and Operational Technologies Convergence tool has inputs and outputs as follows:

Tool inputs, AI-Sim information and operational technologies convergence tool data acquisition can be collected from five main sources: technical data, O&M data, simulation data, procurement data, and economic data. Technical data includes; data on available and used information technology, and data on available and used operational technologies. O&M data includes; real-time operational data of the desalination system under research and development, historical operation and maintenance data of the desalination system under research and development, real-time operational data of other desalination systems, historical operation and maintenance data of other desalination systems, initial operation and maintenance plans. Simulation data includes; data generated from simulation and optimization processes. Procurement data includes; data on equipment available for use with IoT capabilities. Economic data includes; data on measures of benefits from the integration between information technology and operational technologies.

Tool outputs, AI-Sim information and operational technologies convergence tool in demand outputs are prediction, optimization, and recommendation.

Prediction includes; evaluating measurement devices in terms of accuracy, quality, and necessity within the plant, while defining the required specifications based on actual needs. Optimization includes; proposing strategies for maximizing the integration of information technology and operational technology. Recommendation includes; suggesting data analysis programs tailored to the nature of desalination systems and the most suitable operational models, identifying additional data and information required in IT systems to enhance operations, proposing mechanisms to bridge the gap between information technology and operational technologies, and Suggesting improvements for equipment and components to enable compatibility with the IoT, As shown in Figure 78.

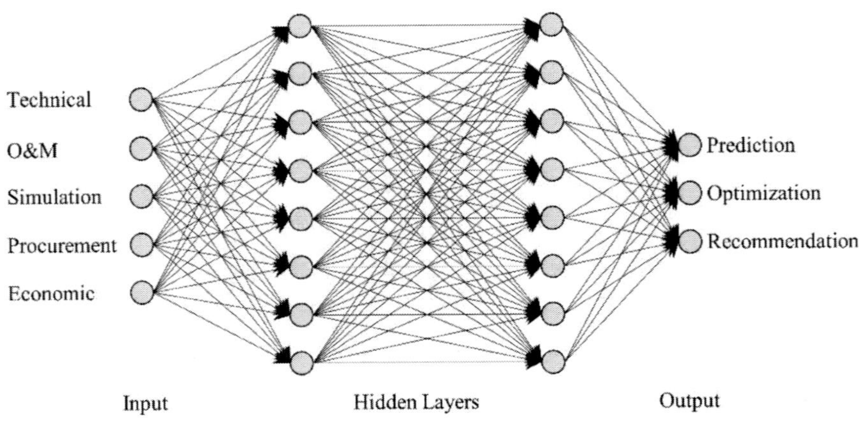

Figure 78. Architecture of AI-Sim Information and Operational Technologies Convergence Tool Model.

AI-Sim IT/OT convergence consists of data integration, data analytics, simulation and modeling, and decision support systems. Using data from different sources, like instrumentation measurements, monitoring and control systems, and enterprise software, to make sense of it all; using data analytics and machine learning algorithms to find insights and guess how the desalination system will behave; simulating and modeling to look at how the system will behave and guess what will happen; and creating a decision support system that helps operators make choices in real time by giving them information and suggestions.

AI-Sim IT/OT convergence tool techniques consist of machine learning algorithms, digital twinning, and simulation software. Machine learning algorithms such as neural networks and decision trees to simulate desalination system behavior and predict outcomes; digital twinning creation for desalination system components and processes to simulate actions and predict outcomes; simulation software: use simulation software to model and simulate desalination system behavior. AI-Sim IT/OT convergence tool architecture consists of a data integration layer, a data analytics layer, a simulation and modeling layer, and a decision support system layer. The data integration layer integrates data from various sources, such as sensors, monitoring and control systems, and enterprise software; the data analytics layer applies data analytics and machine learning algorithms to extract insights and predict desalination system behavior; the simulation and modeling layer uses simulation and modeling techniques to analyze desalination system behavior and predict outcomes; and the decision support system layer develops a decision support system that provides real-time insights and recommendations to operators.

AI-Sim IT/OT convergence offers a powerful tool to improve efficiency by enabling real-time monitoring and predictive maintenance, leading to improved efficiency and reduced downtime; reduced costs by enabling optimized operation and maintenance, leading to reduced energy consumption and lower maintenance costs; and enhanced decision-making by providing real-time insights and recommendations, enabling operators to make informed decisions. KPIs for desalination system IT/OT convergence include efficiency, quality, reliability, safety, environmental, and financial aspects. In consist of efficiency KPIs, such as energy consumption through monitor energy consumption in real-time and optimize energy usage, water production through monitor water production in real-time and optimize production rates, and plant availability through monitor plant availability and downtime to optimize maintenance scheduling; in consist of quality KPIs, such as water quality through monitor water quality in real-time and optimize treatment processes, salinity level through monitor salinity levels in real-time and optimize desalination processes, and pH Level through monitor pH levels in real-time and optimize treatment processes; in consist of reliability KPIs, such as equipment failure rate, downtime, and MTBF to optimize maintenance scheduling; in consist of safety KPIs, such as safety incidents, near misses, and compliance with safety regulations to optimize safety protocols; in consist of environmental KPIs, such as carbon footprint to optimize energy usage, water usage to optimize water conservation, and waste generation to optimize waste reduction; in consist of financial KPIs, such as operating costs to optimize energy usage and maintenance scheduling, capital expenditures to optimize investment decisions, and ROI to optimize investment decisions.

13. Challenges for Using Artificial Intelligence and Simulation for desalination

While the potential of AI in seawater desalination systems optimization is enormous, it's not without some challenges and considerations. Addressing these challenges will be essential for successful adoption of AI and simulation in desalination. Challenges such as; data-related challenges, AI and simulation challenges, domain expertise and knowledge transfer challenges, regulatory and economic challenges, cybersecurity challenges, and organizational and cultural challenges. As shown in Figure 79. Concerns about data in desalination systems involve several essential difficulties, including data quality and availability, data scarcity, integration and interoperability, along with security and privacy. These systems frequently have difficulties in obtaining high-quality and pertinent data, which is crucial for the creation and precision of AI models. Insufficient, inconsistent, or erroneous data can substantially impede the development of dependable AI models, restricting their accuracy and efficacy. Moreover, the absence of essential data for evaluating and training AI systems might hinder the overall efficacy of desalination plants.

The intricacy of desalination systems is exacerbated by the integration of several data sources and platforms, which hinders data integration and interchange. This fragmentation may result in the creation of data silos, hindering the advancement of holistic AI models. Moreover, desalination systems manage sensitive information, such as treatment process specifications and water quality indicators, requiring stringent security protocols to avert unwanted access, cyberattacks, and data breaches. The complexities of desalination processes, marked by non-linear interactions and correlations among factors, hinder the creation of precise AI models. AI systems must possess the ability to recognize intricate linkages and feedback loops to guarantee efficient modeling.

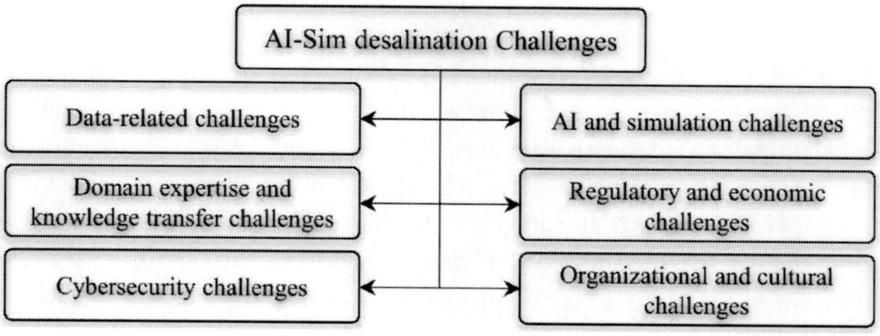

Figure 79. AI and Simulation Challenges for Desalination.

Furthermore, comprehensive desalination simulations demand significant computing resources, requiring AI models to be efficient and scalable to handle massive datasets and intricate simulations. To promote openness and confidence in decision-making, AI systems must deliver clear and explained outcomes. Nonetheless, obstacles concerning domain expertise and information transfer endure, since desalination specialists may be deficient in AI and machine learning ability. This disparity highlights the necessity for collaboration between desalination experts and AI researchers to provide efficient solutions customized for desalination facilities. The effective application of AI in desalination depends significantly on cooperation among industry stakeholders, AI researchers, and desalination specialists. The transfer of knowledge and collaboration can guarantee the proper integration of AI technologies into current systems. The widespread use of AI technologies depends on the standardization and interoperability of AI systems and data formats, which can enable integration, improve data exchange, and foster collaboration within the industry.

Regulatory and economic factors significantly influence the use of AI in desalination. The ongoing development of regulatory frameworks and standards for AI in this industry may create regulatory ambiguity, thereby dissuading facilities from using AI solutions. Furthermore, AI systems must exhibit economic feasibility and a positive ROI to validate their deployment. The public's view and acceptance of AI in desalination can significantly impact its adoption and the development of regulatory frameworks, requiring desalination facilities to address public concerns and effectively convey the advantages of AI technology. Security is a critical problem, as AI systems may be susceptible to data breaches and cyber-attacks. Desalination facilities must establish stringent cybersecurity standards to safeguard data and AI systems, guaranteeing secure data transit and storage through improved security measures, access restrictions, and encryption methods. AI models must be designed with an emphasis on security and robustness, integrating safeguards against data poisoning and manipulation of the models. Ultimately, organizational and cultural obstacles must be confronted for effective AI application. The implementation of AI solutions demands substantial alterations in business culture and procedures, requiring efficient management change tactics. Staff at desalination plants must have sufficient training and knowledge to proficiently run and maintain AI systems.

Conclusion

The integration of AI-Sim technology in desalination facilities is a feasible solution to challenges associated with high energy usage, maintenance costs, and variations in water quality. AI-Sim improves operational efficiency, promotes skill development among operators, and improves system sustainability through sophisticated data analytics, predictive modeling, and simulation capabilities. The tools and framework developed by AI-Sim are essential for enhancing desalination operations, facilitating real-time performance enhancements and compliance with regulations. The comprehensive system reduces costs and environmental impact while enhancing the reliability and quality of water supplies, so facilitating more effective water management in light of increasing worldwide needs. This comprehensive strategy guarantees that water delivery systems are both efficient and robust, capable of adjusting to evolving conditions and needs. By utilizing advanced technology, stakeholders may make educated decisions that enhance long-term sustainability and security of water resources.

References

[1] M. A. Iqbal et al., "*Advanced photocatalysis as a viable and sustainable wastewater treatment process: A comprehensive review*," p. 118947, 2024.

[2] S. M. J. Seyed Sabour and B. Ghorashi, "A comprehensive review of major water desalination techniques and mineral extraction from saline water," *Separation and Purification Technology,* vol. 349, p. 127913, 2024/12/03/ 2024.

[3] J. Eke, A. Yusuf, A. Giwa, and A. J. D. Sodiq, "*The global status of desalination: An assessment of current desalination technologies, plants and capacity,*" vol. 495, p. 114633, 2020.

[4] S. Almasoudi and B. J. S. C. O. W. Jamoussi, "*Desalination technologies and their environmental impacts: a review,*" vol. 1, p. 100002, 2024.

[5] E. Jones, M. Qadir, M. T. Van Vliet, V. Smakhtin, and S.-m. J. S. o. t. T. E. Kang, "*The state of desalination and brine production: A global outlook,*" vol. 657, pp. 1343-1356, 2019.

[6] M. Nair, D. J. D. Kumar, and W. Treatment, "*Water desalination and challenges: The Middle East perspective: a review,*" vol. 51, no. 10-12, pp. 2030-2040, 2013.

[7] A. N. Angelakis, S. V. J. W. S. Spyridakis, and T. W. Supply, "*Major urban water and wastewater systems in Minoan Crete, Greece,*" vol. 13, no. 3, pp. 564-573, 2013.

[8] M. J. A. i. C. C. Costa, Global Warming, Biological Problems and N. Hazards, "*The Climate and History of Algarve—Interactions,*" pp. 70-75, 2010.

[9] A. N. Angelakis, D. J. W. S. Spyridakis, and T. W. Supply, "*A brief history of water supply and wastewater management in ancient Greece,*" vol. 10, no. 4, pp. 618-628, 2010.

[10] A. J. H. o. c. Delyannis and t. Athens, "*Introduction to chemical technology,*" vol. 410, 1960.

[11] S. Zhu, L. Xue, and Z. J. T. W. T. Xu, "*The development history and an analysis of status quo of desalination at home and abroad,*" vol. 40, pp. 12-23, 2014.

[12] S. A. J. P. i. e. Kalogirou and c. science, "*Seawater desalination using renewable energy sources,*" vol. 31, no. 3, pp. 242-281, 2005.

[13] A. Angelakis et al., "*Desalination: From Ancient to Present and Future. Water 2021, 13, 2222,*" ed: s Note: MDPI stays neutral with regard to jurisdictional claims in published …, 2021.

[14] J. Needham, *Science and civilisation in China.* Cambridge University Press, 1974.

[15] T. Hinkebein, "Desalination: Limitations and challenges," in *Water and sustainable development: Opportunities for the chemical sciences: A workshop*

report to the chemical sciences roundtable, 2004, pp. 29-39: National Academies Press.

[16] D. J. I. T. A. o. h. w. a. o. w. h. w. i.-t. c. b. p. b.-.-b.-m.-d. Talbot, "*Bankrolling the 10 Breakthrough Technologies: Megascale Desalination*," 2016.

[17] L. J. D. Garcia-Rodriguez, "*Seawater desalination driven by renewable energies: a review,*" vol. 143, no. 2, pp. 103-113, 2002.

[18] H. Cooley, P. H. Gleick, and G. H. Wolff, "*Desalination, with a grain of salt: a California perspective*," ed: Pacific Institute for Studies in Development, Environment, and Security …, 2006.

[19] M. S. Singleton, G. Heiss, A. J. P. R. E. S. Hübler, Nonlinear, and S. M. Physics, "*Optimization of ramified absorber networks doing desalination*," vol. 83, no. 1, p. 016308, 2011.

[20] E. Koutroulis and D. J. D. Kolokotsa, "*Design optimization of desalination systems power-supplied by PV and W/G energy sources*," vol. 258, no. 1-3, pp. 171-181, 2010.

[21] T. M. Missimer and R. G. J. D. Maliva, "*Environmental issues in seawater reverse osmosis desalination: Intakes and outfalls*," vol. 434, pp. 198-215, 2018.

[22] T. M. Missimer and R. G. Maliva, "Environmental issues in seawater reverse osmosis desalination: Intakes and outfalls," *Desalination,* vol. 434, pp. 198-215, 2018/05/15/ 2018.

[23] N. Ghaffour, T. M. Missimer, and G. L. J. D. Amy, "*Technical review and evaluation of the economics of water desalination: Current and future challenges for better water supply sustainability*," vol. 309, pp. 197-207, 2013.

[24] T. Pankratz, "Overview of intake systems for seawater reverse osmosis facilities," in *Intakes and Outfalls for Seawater Reverse-Osmosis Desalination Facilities: Innovations and Environmental Impacts*: Springer, 2015, pp. 3-17.

[25] F. J. A. W. S. Esmaeilion, "*Hybrid renewable energy systems for desalination*," vol. 10, no. 3, p. 84, 2020.

[26] F. E. Ahmed, R. Hashaikeh, and N. J. D. Hilal, "*Hybrid technologies: The future of energy efficient desalination–A review*," vol. 495, p. 114659, 2020.

[27] H. Nassrullah, S. F. Anis, R. Hashaikeh, and N. J. D. Hilal, "*Energy for desalination: A state-of-the-art review*," vol. 491, p. 114569, 2020.

[28] V. G. Gude and V. J. P. i. E. Fthenakis, "*Energy efficiency and renewable energy utilization in desalination systems,*" vol. 2, no. 2, p. 022003, 2020.

[29] M. A. A. Shahmirzadi, S. S. Hosseini, J. Luo, and I. J. J. o. e. m. Ortiz, "*Significance, evolution and recent advances in adsorption technology, materials and processes for desalination, water softening and salt removal*," vol. 215, pp. 324-344, 2018.

[30] K. Elsaid, E. T. Sayed, M. A. Abdelkareem, M. S. Mahmoud, M. Ramadan, and A. J. J. o. E. C. E. Olabi, "*Environmental impact of emerging desalination technologies: A preliminary evaluation*," vol. 8, no. 5, p. 104099, 2020.

[31] D. M. J. D. Al-Gobaisi, "*Conceptual specification for improved automation and total process care in large-scale desalination plants of the future*," vol. 95, no. 3, pp. 287-297, 1994.

[32] S. Honarparvar, X. Zhang, T. Chen, C. Na, and D. J. C. O. i. C. E. Reible, "*Modeling technologies for desalination of brackish water—toward a sustainable water supply,*" vol. 26, pp. 104-111, 2019.

[33] G. F. Clark et al., "*First large-scale ecological impact study of desalination outfall reveals trade-offs in effects of hypersalinity and hydrodynamics*," vol. 145, pp. 757-768, 2018.

[34] E. Pita, P. Pita, M. Sánchez-Barriga, L. Martín, J. J. D. Olalde, and W. Treatment, "*Design of marine outfalls for reducing environmental impact of brine*," vol. 259, pp. 300-307, 2022.

[35] R. Bashitialshaaer, K. M. Persson, and M. Larson, "New criteria for brine discharge outfalls from desalination plants," in *Intakes and Outfalls for Seawater Reverse-Osmosis Desalination Facilities: Innovations and Environmental Impacts*, 2015, pp. 451-467: Springer.

[36] A. Panagopoulos, K.-J. Haralambous, and M. J. S. o. t. T. E. Loizidou, "*Desalination brine disposal methods and treatment technologies-A review*," vol. 693, p. 133545, 2019.

[37] T. J. J. o. c. w. r. Younos and education, "*Environmental issues of desalination,*" vol. 132, no. 1, p. 3, 2005.

[38] J. R. J. W. R. M. Ziolkowska, "*Is desalination affordable?—regional cost and price analysis*," vol. 29, pp. 1385-1397, 2015.

[39] H. K. Abdulrahim et al., "*Reverse osmosis desalination system and algal blooms Part III: SWRO pretreatment*," vol. 60, pp. 11-38, 2017.

[40] J. S. J. D. Chang, "*Understanding the role of ecological indicator use in assessing the effects of desalination plants*," vol. 365, pp. 416-433, 2015.

[41] A. Valipour, N. Hamnabard, K.-S. Woo, and Y.-H. J. J. o. E. M. Ahn, "*Performance of high-rate constructed phytoremediation process with attached growth for domestic wastewater treatment: Effect of high TDS and Cu*," vol. 145, pp. 1-8, 2014.

[42] C. M. Hobbs, J. M. Arevalo, and C. A. J. O. Kiefer, "*Concentrate management: weigh benefits and drawbacks to ensure success*," vol. 42, no. 9, pp. 22-25, 2016.

[43] P. Amoatey et al., "A critical review of environmental and public health impacts from the activities of evaporation ponds," vol. 796, p. 149065, 2021.

[44] A. Shokri and M. Sanavi Fard, "Techno-economic assessment of water desalination: Future outlooks and challenges," *Process Safety and Environmental Protection,* vol. 169, pp. 564-578, 2023/01/01/ 2023.

[45] T. Mezher, H. Fath, Z. Abbas, and A. Khaled, "Techno-economic assessment and environmental impacts of desalination technologies," *Desalination,* vol. 266, no. 1, pp. 263-273, 2011/01/31/ 2011.

[46] J. R. Ziolkowska, "Is Desalination Affordable?—Regional Cost and Price Analysis," *Water Resources Management,* vol. 29, no. 5, pp. 1385-1397, 2015/03/01 2015.

[47] A. Papangelou, M. Brenda, V. Hirsch, S. Liehr, and W. J. I. W. R. M. I. W.-S. R. Urban, "*Economic viability and financing options*," p. 200, 2018.

[48] Y. Cai, J. Wu, S. Q. Shi, J. Li, and K.-H. Kim, "Advances in desalination technology and its environmental and economic assessment," *Journal of Cleaner Production,* vol. 397, p. 136498, 2023/04/15/ 2023.

[49] A. Shokri and M. Sanavi Fard, "A sustainable approach in water desalination with the integration of renewable energy sources: Environmental engineering challenges and perspectives," *Environmental Advances,* vol. 9, p. 100281, 2022/10/01/ 2022.

[50] A. R. Hurtado and J. J. L. Berbel, "*A Cost–Benefit Analysis of Reclaimed Water and Desalinated Seawater for Irrigation in Axarquia, Southern Spain*," vol. 13, no. 12, p. 2156, 2024.

[51] C. Li, "*Economic and cost-effectiveness analysis of integrated desalination and brine treatment systems*," 2022.

[52] F. E. Ahmed, A. Khalil, and N. J. D. Hilal, "*Emerging desalination technologies: Current status, challenges and future trends,*" vol. 517, p. 115183, 2021.

[53] M. Walschot, P. Luis, and M. J. W. I. Liégeois, "*The challenges of reverse osmosis desalination: Solutions in Jordan,*" vol. 45, no. 2, pp. 112-124, 2020.

[54] M. Schorr, *Desalination: Trends and technologies*. BoD–Books on Demand, 2011.

[55] P. Moilanen and U.-M. Mroueh, "Mobilising funding in the water sector: The potential for private sector participation and desalination in the levant region," in *Managing blue gold: New perspectives on water security in the Levantine Middle East*: Ulkopoliittinen instituutti, 2010, pp. 109-125.

[56] M. Taylor, K. Daniel, A. Ilas, and E. Y. J. I. R. E. A. M. C. So, Abu Dhabi, UAE, "*Renewable power generation costs in 2014,*" 2015.

[57] A. M. Ghaithan, A. Al-Hanbali, A. Mohammed, A. M. Attia, H. Saleh, and O. J. R. E. Alsawafy, "*Optimization of a solar-wind-grid powered desalination system in Saudi Arabia,*" vol. 178, pp. 295-306, 2021.

[58] H. M. S. Amr Mohamed Seoudy, "Reverse osmosis plants environmental performance index," presented a*t the Desalination for the Environment Clean Water and Energy*, Limassol, Cyprus 22–26 May 2023.

[59] M. Benaissa, O. Rouane-Hacene, Z. Boutiba, D. Habib, M. E. Guibbolini-Sabatier, and C. J. R. S. i. M. S. Risso-De Faverney, "*Ecotoxicological effects assessment of brine discharge from desalination reverse osmosis plant in Algeria (South Western Mediterranean)*," vol. 39, p. 101407, 2020.

[60] A. Belatoui, H. Bouabessalam, O. R. Hacene, J. A. de-la-Ossa-Carretero, E. Martinez-Garcia, and J. L. Sanchez-Lizaso, "Environmental effects of brine discharge from two desalination plants in Algeria (South Western Mediterranean)," *Desalination and Water Treatment,* vol. 76, pp. 311-318, 2017/05/01/ 2017.

[61] S. Lattemann and T. J. D. Höpner, "*Environmental impact and impact assessment of seawater desalination*," vol. 220, no. 1-3, pp. 1-15, 2008.

[62] A. Bazargan, *A multidisciplinary introduction to desalination.* CRC Press, 2022.

[63] F. J. E. S. Alshahri and P. Research, "*Heavy metal contamination in sand and sediments near to disposal site of reject brine from desalination plant, Arabian Gulf: Assessment of environmental pollution,*" vol. 24, no. 2, pp. 1821-1831, 2017.

[64] M. Omerspahic, H. Al-Jabri, S. A. Siddiqui, and I. J. F. i. M. S. Saadaoui, "*Characteristics of desalination brine and its impacts on marine chemistry and*

health, with emphasis on the Persian/Arabian gulf: a review," vol. 9, p. 845113, 2022.

[65] R. Sirota et al., "*Impacts of desalination brine discharge on benthic ecosystems,*" vol. 58, no. 13, pp. 5631-5645, 2024.

[66] K. Elsaid, M. Kamil, E. T. Sayed, M. A. Abdelkareem, T. Wilberforce, and A. Olabi, "Environmental impact of desalination technologies: A review," *Science of The Total Environment,* vol. 748, p. 141528, 2020/12/15/ 2020.

[67] H.-M.-S. Amr-Mohamed-Seoudy, "Desalination in Egypt Between the Present and the Future, Industry or Trade," presented at *The International Desalination Association World Congress on CROSSROADS TO SUSTAINABILITY 2019, Dubai, UAE, 20 – 24 October 2019.*

[68] A. Seoudy, "Socioeconomic Effects of Public Private Partnership Desalination Projects in Egypt," presented at the IDRA 2024 World Congress -Addressing Water Scarcity, Abo Dabi, December 2024.

[69] E. Brynjolfsson and S. J. A. i. c. Yang, "*Information technology and productivity: a review of the literature,*" vol. 43, pp. 179-214, 1996.

[70] D. G. J. G. Rossiter, "*Past, present & future of information technology in pedometrics,*" vol. 324, pp. 131-137, 2018.

[71] M. Javaid, A. Haleem, R. P. Singh, and A. K. Sinha, "Digital economy to improve the culture of industry 4.0: A study on features, implementation and challenges. Green Technologies and Sustainability, 2 (2)," ed, 2024.

[72] J. A. Younis, M. Raouf, M. A. J. J. o. P. P. Abdallah, and Wellbeing, "The Influence of SCADA Information Technology System Application on Improving Performance Efficiency (Field study on the Ministry of Electricity and Oil of the Kurdistan Region of Iraq)," vol. 6, no. 2, pp. 1040-1061, 2022.

[73] J. Eke, A. Yusuf, A. Giwa, and A. Sodiq, "The global status of desalination: An assessment of current desalination technologies, plants and capacity," *Desalination,* vol. 495, p. 114633, 2020/12/01/ 2020.

[74] E. Jones, M. Qadir, M. T. H. van Vliet, V. Smakhtin, and S.-m. Kang, "The state of desalination and brine production: A global outlook," *Science of The Total Environment,* vol. 657, pp. 1343-1356, 2019/03/20/ 2019.

[75] M. Papapetrou, A. Cipollina, U. La Commare, G. Micale, G. Zaragoza, and G. Kosmadakis, "Assessment of methodologies and data used to calculate desalination costs," *Desalination,* vol. 419, pp. 8-19, 2017/10/01/ 2017.

[76] M. K. Wittholz, B. K. O'Neill, C. B. Colby, and D. Lewis, "Estimating the cost of desalination plants using a cost database," *Desalination,* vol. 229, no. 1, pp. 10-20, 2008/09/15/ 2008.

[77] A. Lamei, P. van der Zaag, and E. von Münch, "Basic cost equations to estimate unit production costs for RO desalination and long-distance piping to supply water to tourism-dominated arid coastal regions of Egypt," *Desalination,* vol. 225, no. 1, pp. 1-12, 2008/05/01/ 2008.

[78] K. V. Reddy and N. Ghaffour, "Overview of the cost of desalinated water and costing methodologies," *Desalination,* vol. 205, no. 1, pp. 340-353, 2007/02/05/ 2007.

[79] N. Ghaffour, T. M. Missimer, and G. L. Amy, "Technical review and evaluation of the economics of water desalination: Current and future challenges for better water supply sustainability," *Desalination,* vol. 309, pp. 197-207, 2013/01/15/ 2013.

[80] R. A. Greer, "*A review of public water infrastructure financing in the United States,*" vol. 7, no. 5, p. e1472, 2020.

[81] R. A. Greer, K. Lee, A. Fencl, and G. Sneegas, "Public–Private Partnerships in the Water Sector: The Case of Desalination," *Water Resources Management,* vol. 35, no. 11, pp. 3497-3511, 2021/09/01 2021.

[82] A. Seoudy, "Perspective Chapter: Desalination System Performance Index and Artificial Intelligence," in *Advances in Desalination Insights*, H. Xu, X. Gu, Q. Li, H. Wang, and Y. Sun, Eds. Rijeka: IntechOpen, 2024.

[83] A. Shokri and M. Sanavi Fard, "A comprehensive overview of environmental footprints of water desalination and alleviation strategies," *International Journal of Environmental Science and Technology,* vol. 20, no. 2, pp. 2347-2374, 2023/02/01 2023.

[84] M. Nasrollahi, A. Motevali, A. Banakar, and M. Montazeri, "Comparison of environmental impact on various desalination technologies," *Desalination,* vol. 547, p. 116253, 2023/02/01/ 2023.

[85] P. H. Gomes, S. P. Pereira, T. C. L. Tavares, T. M. Garcia, and M. O. Soares, "Impacts of desalination discharges on phytoplankton and zooplankton: Perspectives on current knowledge," *Science of The Total Environment,* vol. 863, p. 160671, 2023/03/10/ 2023.

[86] S. Almasoudi and B. Jamoussi, "Desalination technologies and their environmental impacts: A review," *Sustainable Chemistry One World,* vol. 1, p. 100002, 2024/03/01/ 2024.

[87] N. Kress, "Chapter 4 - Theoretical Analysis of the Potential Impacts of Desalination on the Marine Environment," in *Marine Impacts of Seawater Desalination*, N. Kress, Ed.: Elsevier, 2019, pp. 53-69.

[88] A. Alvez, D. Aitken, D. Rivera, M. Vergara, N. McIntyre, and F. Concha, "At the crossroads: can desalination be a suitable public policy solution to address water scarcity in Chile's mining zones?," *Journal of Environmental Management,* vol. 258, p. 110039, 2020/03/15/ 2020.

[89] M. Sharifinia, Z. Afshari Bahmanbeigloo, W. O. Smith Jr, C. K. Yap, and M. J. G. c. b. Keshavarzifard, "*Prevention is better than cure: Persian Gulf biodiversity vulnerability to the impacts of desalination plants*," vol. 25, no. 12, pp. 4022-4033, 2019.

[90] L. Wang, C. Violet, R. M. DuChanois, and M. Elimelech, "Derivation of the Theoretical Minimum Energy of Separation of Desalination Processes," *Journal of Chemical Education,* vol. 97, no. 12, pp. 4361-4369, 2020/12/08 2020.

[91] H. Wang and X. Jiang, "Legal measures to reduce marine environmental risks of seawater desalination in China," in *Natural Resources Forum*, 2020, vol. 44, no. 2, pp. 129-143: Wiley Online Library.

[92] L. Henthorne and B. Boysen, "State-of-the-art of reverse osmosis desalination pretreatment," *Desalination,* vol. 356, pp. 129-139, 2015/01/15/ 2015.

[93] Y. Xing, Q. Liu, M. Zhang, Y. Zhen, T. Mi, and Z. Yu, "Effects of temperature and salinity on the asexual reproduction of Aurelia coerulea polyps," *Journal of Oceanology and Limnology,* vol. 38, no. 1, pp. 133-142, 2020/01/01 2020.

[94] M. L. Cambridge, A. Zavala-Perez, G. R. Cawthray, J. Statton, J. Mondon, and G. A. Kendrick, "Effects of desalination brine and seawater with the same elevated salinity on growth, physiology and seedling development of the seagrass Posidonia australis," *Marine Pollution Bulletin,* vol. 140, pp. 462-471, 2019/03/01/ 2019.

[95] C. Kenigsberg, S. Abramovich, and O. J. P. O. Hyams-Kaphzan, "*The effect of long-term brine discharge from desalination plants on benthic foraminifera,*" vol. 15, no. 1, p. e0227589, 2020.

[96] K. L. Petersen et al., "Impact of brine and antiscalants on reef-building corals in the Gulf of Aqaba – Potential effects from desalination plants," *Water Research,* vol. 144, pp. 183-191, 2018/11/01/ 2018.

[97] G. F. Clark et al., "First large-scale ecological impact study of desalination outfall reveals trade-offs in effects of hypersalinity and hydrodynamics," *Water Research,* vol. 145, pp. 757-768, 2018/11/15/ 2018.

[98] V. G. Gude, "Desalination and sustainability – An appraisal and current perspective," *Water Research,* vol. 89, pp. 87-106, 2016/02/01/ 2016.

[99] A. Panagopoulos, K.-J. Haralambous, and M. Loizidou, "Desalination brine disposal methods and treatment technologies - A review," *Science of The Total Environment,* vol. 693, p. 133545, 2019/11/25/ 2019.

[100] G. Amy et al., "Membrane-based seawater desalination: Present and future prospects," *Desalination,* vol. 401, pp. 16-21, 2017/01/02/ 2017.

[101] K. Elsaid, E. T. Sayed, M. A. Abdelkareem, M. S. Mahmoud, M. Ramadan, and A. G. Olabi, "Environmental impact of emerging desalination technologies: A preliminary evaluation," *Journal of Environmental Chemical Engineering,* vol. 8, no. 5, p. 104099, 2020/10/01/ 2020.

[102] I. Ihsanullah, M. A. Atieh, M. Sajid, and M. K. Nazal, "Desalination and environment: A critical analysis of impacts, mitigation strategies, and greener desalination technologies," *Science of The Total Environment,* vol. 780, p. 146585, 2021/08/01/ 2021.

[103] A. Seoudy, "*Perspective Chapter: Desalination System Performance Index and Artificial Intelligence,*" 2024.

[104] F. Ameen, J. A. Stagner, and D. S.-K. J. I. J. o. E. S. Ting, "*The carbon footprint and environmental impact assessment of desalination,*" vol. 75, no. 1, pp. 45-58, 2018.

[105] X. Wei, Z. M. Binger, A. Achilli, K. T. Sanders, and A. E. J. W. r. Childress, "*A modeling framework to evaluate blending of seawater and treated wastewater streams for synergistic desalination and potable reuse,*" vol. 170, p. 115282, 2020.

[106] M. Heihsel, M. Lenzen, A. Malik, and A. J. D. Geschke, "*The carbon footprint of desalination: an input-output analysis of seawater reverse osmosis desalination in Australia for 2005–2015,*" vol. 454, pp. 71-81, 2019.

[107] M. Tajik Mansouri, M. Amidpour, and J. M. Ponce-Ortega, "Optimal integration of organic Rankine cycle and desalination systems with industrial processes:

Energy-water-environment nexus," *Applied Thermal Engineering,* vol. 158, p. 113740, 2019/07/25/ 2019.

[108] V. S. Saji, A. A. Meroufel, and A. A. Sorour, "*Corrosion and fouling control in desalination industry*," 2020.

[109] Q. He, H. Zheng, X. Ma, L. Wang, H. Kong, and Z. Zhu, "Artificial intelligence application in a renewable energy-driven desalination system: A critical review," *Energy and AI,* vol. 7, p. 100123, 2022/01/01/ 2022.

[110] E. T. Sayed et al., "*Application of artificial intelligence techniques for modeling, optimizing, and controlling desalination systems powered by renewable energy resources,*" vol. 413, p. 137486, 2023.

[111] G. C. de Oliveira, E. Bertone, and R. A. J. A. E. Stewart, "*Optimisation modelling tools and solving techniques for integrated precinct-scale energy–water system planning,*" vol. 318, p. 119190, 2022.

[112] H. Sayed, T. A. El-Sayed, M. I. Friswell, and H. H. J. N. D. El-Mongy, "*Nonlinear dynamics analysis of hydraulic turbochargers in reverse osmosis desalination plants*," vol. 112, no. 21, pp. 18817-18846, 2024.

[113] M. J. Blondin, J. Sanchis Sáez, P. M. J. C. I. Pardalos, and O. M. f. C. Engineering, "*Control engineering from classical to intelligent control theory—an overview*," pp. 1-30, 2019.

[114] L. Hu, G. Zhang, Y. Liu, T. Guo, L. Shao, and L. J. J. o. M. C. C. Ying, "*Efficient dendrimers based on naphthalene indenofluorene for two-photon fluorescent imaging in living cells and tissues*," vol. 8, no. 6, pp. 2160-2170, 2020.

[115] H. M. El-Arwash, A. M. Azmy, and E. M. Rashad, "A GA-based initialization of PSO for optimal APFS allocation in water desalination plant," in *2017 Nineteenth International Middle East Power Systems Conference (MEPCON)*, 2017, pp. 1378-1384: IEEE.

[116] B. Chen, C. Vondrick, and H. J. S. R. Lipson, "*Visual behavior modelling for robotic theory of mind*," vol. 11, no. 1, p. 424, 2021.

[117] Q. Sun, L. J. C. J. o. P. Yang, and E. Systems, "*From independence to interconnection—A review of AI technology applied in energy systems,*" vol. 5, no. 1, pp. 21-34, 2019.

[118] Y. K. Dwivedi et al., "*Artificial Intelligence (AI): Multidisciplinary perspectives on emerging challenges, opportunities, and agenda for research, practice and policy,*" vol. 57, p. 101994, 2021.

[119] J. Hays and A. A. J. A. T. o. g. Efros, "*Scene completion using millions of photographs*," vol. 26, no. 3, pp. 4-es, 2007.

[120] J. H. Thrall et al., "*Artificial intelligence and machine learning in radiology: opportunities, challenges, pitfalls, and criteria for success*," vol. 15, no. 3, pp. 504-508, 2018.

[121] A. Trunk, H. Birkel, and E. J. B. R. Hartmann, "*On the current state of combining human and artificial intelligence for strategic organizational decision making*," vol. 13, no. 3, pp. 875-919, 2020.

[122] Z. Jan et al., "*Artificial intelligence for industry 4.0: Systematic review of applications, challenges, and opportunities*," vol. 216, p. 119456, 2023.

[123] M. Soori, B. Arezoo, and R. J. C. R. Dastres, "*Artificial intelligence, machine learning and deep learning in advanced robotics, a review*," vol. 3, pp. 54-70, 2023.

[124] M. Pellicelli, "Chapter five - Managing the supply chain: technologies for digitalization solutions," in *The Digital Transformation of Supply Chain Management*, M. Pellicelli, Ed.: Elsevier, 2023, pp. 101-152.

[125] O. J. D. A. I. Kaynak, "T*he golden age of artificial intelligence: Inaugural editorial*," vol. 1, ed: Springer, 2021, pp. 1-7.

[126] J. J. a. p. a. Schmidhuber, "Annotated history of modern ai and deep learning," 2022.

[127] J. B. Krieger, F. Bouder, M. Wibral, and R. J. J. o. R. R. Almeida, "*A systematic literature review on risk perception of Artificial Narrow Intelligence*," pp. 1-19, 2024.

[128] O. Arshi and A. Chaudhary, "Overview of Artificial General Intelligence (AGI)," in *Artificial General Intelligence (AGI) Security: Smart Applications and Sustainable Technologies*: Springer, 2024, pp. 1-26.

[129] K. M. Stiefel and J. S. J. F. i. A. I. Coggan, "*The energy challenges of artificial superintelligence*," vol. 6, p. 1240653, 2023.

[130] S. B. Jige and J. J. Launa, "Role of Artificial Intelligence in Sustainable Development," in *Achieving Sustainability with AI Technologies*: IGI Global Scientific Publishing, 2025, pp. 311-326.

[131] N. Ameen, G. D. Sharma, S. Tarba, A. Rao, R. J. P. Chopra, and marketing, "*Toward advancing theory on creativity in marketing and artificial intelligence*," vol. 39, no. 9, pp. 1802-1825, 2022.

[132] S. J. E. I. Dolgikh, "*Self-awareness in natural and artificial intelligent systems: a unified information-based approach*," vol. 17, no. 5, pp. 4095-4114, 2024.

[133] Y. Xu, H. Mellmann, and H.-D. Burkhard, "An approach to close the gap between simulation and real robots," in *International Conference on Simulation, Modeling, and Programming for Autonomous Robots*, 2010, pp. 533-544: Springer.

[134] Y. Tang, L. Li, X. J. C. s. m. Liu, and simulation, "*State-of-the-art development of complex systems and their simulation methods*," vol. 1, no. 4, pp. 271-290, 2021.

[135] S. Spolaor, M. S. Nobile, G. Mauri, P. Cazzaniga, and D. J. I. T. o. F. S. Besozzi, "*Coupling mechanistic approaches and fuzzy logic to model and simulate complex systems*," vol. 28, no. 8, pp. 1748-1759, 2019.

[136] A. Greasley and J. S. J. J. o. t. O. R. S. Edwards, "*Enhancing discrete-event simulation with big data analytics: A review,*" vol. 72, no. 2, pp. 247-267, 2021.

[137] H. Smith, "Chapter 9 - Process Simulation and Modeling," in *Gold Ore Processing (Second Edition)*, M. D. Adams, Ed.: Elsevier, 2016, pp. 141-148.

[138] J. Crompton, "5 - Data management from the DCS to the historian and HMI," in *Machine Learning and Data Science in the Power Generation Industry*, P. Bangert, Ed.: Elsevier, 2021, pp. 93-122.

[139] T. R. Wanasinghe, M. Galagedarage Don, R. Arunthavanathan, and R. G. Gosine, "Chapter Four - Industry 4.0 based process data analytics platform," in *Methods in Chemical Process Safety*, vol. 6, F. Khan, H. Pasman, and M. Yang, Eds.: Elsevier, 2022, pp. 101-137.

[140] E. Badakhshan and P. J. O. R. P. Ball, "*A simulation-optimization approach for integrating physical and financial flows in a supply chain under economic uncertainty*," vol. 10, p. 100270, 2023.

[141] W. Cai et al., "*Analysis of heat extraction performance and long-term sustainability for multiple deep borehole heat exchanger array: A project-based study*," vol. 289, p. 116590, 2021.

[142] A. Koulouris, N. Misailidis, D. J. F. Petrides, and B. Processing, "*Applications of process and digital twin models for production simulation and scheduling in the manufacturing of food ingredients and products*," vol. 126, pp. 317-333, 2021.

[143] F. Gielnik, O. Kinas, and T. Leibfried, "Modeling and Dynamic Simulation of Non-standard Operating Conditions in Low-Voltage Grids Considering Different Network Topologies," in *2023 58th International Universities Power Engineering Conference (UPEC)*, 2023, pp. 1-6: IEEE.

[144] S. A. Samaras, C. L. Adkins, and C. D. J. J. o. E. f. B. White, "*Developing critical thinking skills: Simulations vs. cases*," vol. 97, no. 4, pp. 270-276, 2022.

[145] P. A. Langley, "Beyond Just Talking Strategy: Using Gaming Simulations to Catalyze Airline Managers' Buy-in to Novel Strategies that Can Shape or Adapt to Profit Cyclicality," *Systemic Practice and Action Research,* vol. 37, no. 2, pp. 187-205, 2024/04/01 2024.

[146] M. Banković et al., "Teaching graduate students how to review research articles and respond to reviewer comments," in *Advances in computers*, vol. 116no. 1): Elsevier, 2020, pp. 1-63.

[147] R. P. França, A. C. B. Monteiro, R. Arthur, and Y. J. T. i. d. l. m. Iano, "*An overview of deep learning in big data, image, and signal processing in the modern digital age*," pp. 63-87, 2021.

[148] A. B. Rashid and A. K. J. H. A. Kausik, "*AI revolutionizing industries worldwide: A comprehensive overview of its diverse applications*," p. 100277, 2024.

[149] A. Kumbhar, P. G. Dhawale, S. Kumbhar, U. Patil, and P. Magdum, "A comprehensive review: Machine learning and its application in integrated power system," *Energy Reports,* vol. 7, pp. 5467-5474, 2021/11/01/ 2021.

[150] J. M. Duarte and L. J. A. i. r. Berton, "*A review of semi-supervised learning for text classification*," vol. 56, no. 9, pp. 9401-9469, 2023.

[151] A. S. Kwekha-Rashid, H. N. Abduljabbar, and B. Alhayani, "Coronavirus disease (COVID-19) cases analysis using machine-learning applications," *Applied Nanoscience,* vol. 13, no. 3, pp. 2013-2025, 2023/03/01 2023.

[152] E. F. Morales and H. J. Escalante, "Chapter 6 - A brief introduction to supervised, unsupervised, and reinforcement learning," in *Biosignal Processing and Classification Using Computational Learning and Intelligence*, A. A. Torres-García, C. A. Reyes-García, L. Villaseñor-Pineda, and O. Mendoza-Montoya, Eds.: Academic Press, 2022, pp. 111-129.

[153] N. Le, V. S. Rathour, K. Yamazaki, K. Luu, and M. Savvides, "Deep reinforcement learning in computer vision: a comprehensive survey," *Artificial Intelligence Review,* vol. 55, no. 4, pp. 2733-2819, 2022/04/01 2022.

[154] M. Batool, O. Sanumi, and J. Jankovic, "Application of artificial intelligence in the materials science, with a special focus on fuel cells and electrolyzers," *Energy and AI,* vol. 18, p. 100424, 2024/12/01/ 2024.

[155] M. G. Kibria, K. Nguyen, G. P. Villardi, O. Zhao, K. Ishizu, and F. Kojima, "Big Data Analytics, Machine Learning, and Artificial Intelligence in Next-Generation Wireless Networks," *IEEE Access,* vol. 6, pp. 32328-32338, 2018.

[156] E. Blasch et al., "Machine Learning/Artificial Intelligence for Sensor Data Fusion–Opportunities and Challenges," *IEEE Aerospace and Electronic Systems Magazine,* vol. 36, no. 7, pp. 80-93, 2021.

[157] I. Pan, L. R. Mason, and O. K. Matar, "Data-centric Engineering: integrating simulation, machine learning and statistics. Challenges and opportunities," *Chemical Engineering Science,* vol. 249, p. 117271, 2022/02/15/ 2022.

[158] K. Duraisamy, "Perspectives on machine learning-augmented Reynolds-averaged and large eddy simulation models of turbulence," *Physical Review Fluids,* vol. 6, no. 5, p. 050504, 05/12/ 2021.

[159] B. Saravi et al., "*Artificial Intelligence-Driven Prediction Modeling and Decision Making in Spine Surgery Using Hybrid Machine Learning Models*," vol. 12, no. 4, p. 509, 2022.

[160] C. P. Dai and F. Ke, "Educational applications of artificial intelligence in simulation-based learning: A systematic mapping review," *Computers and Education: Artificial Intelligence,* vol. 3, p. 100087, 2022/01/01/ 2022.

[161] S. Łaniewski and R. J. a. p. a. Ślepaczuk, "Enhancing literature review with LLM and NLP methods. Algorithmic trading case," 2024.

[162] A. Norouzi, H. Heidarifar, H. Borhan, M. Shahbakhti, and C. R. Koch, "Integrating Machine Learning and Model Predictive Control for automotive applications: A review and future directions," *Engineering Applications of Artificial Intelligence,* vol. 120, p. 105878, 2023/04/01/ 2023.

[163] C. Birchler, S. Khatiri, B. Bosshard, A. Gambi, and S. Panichella, "Machine learning-based test selection for simulation-based testing of self-driving cars software," *Empirical Software Engineering,* vol. 28, no. 3, p. 71, 2023/04/26 2023.

[164] S. Samant et al., "Artificial Intelligence, Computational Simulations, and Extended Reality in Cardiovascular Interventions," *JACC: Cardiovascular Interventions,* vol. 16, no. 20, pp. 2479-2497, 2023/10/23 2023.

[165] J. Duan, S. Yu, H. L. Tan, H. Zhu, and C. Tan, "A Survey of Embodied AI: From Simulators to Research Tasks," *IEEE Transactions on Emerging Topics in Computational Intelligence,* vol. 6, no. 2, pp. 230-244, 2022.

[166] G. Deepak, M. Parthiban, S. S. Nath, B. Sulaiman Alfurhood, B. Mouleswararao, and V. Ravi Kishore, "Ai-enhanced thermal modeling for integrated process-product-system optimization in zero-defect manufacturing chains," *Thermal Science and Engineering Progress,* vol. 55, p. 102945, 2024/10/01/ 2024.

[167] M. Shafiq, K. Thakre, R. Pandurangan, and R. V. S. Lalitha, "Generative AI designs the next generation of smart materials from pixels to products," *The International Journal of Advanced Manufacturing Technology,* 2025/01/14 2025.

[168] S. Chinchanikar and A. A. Shaikh, "A Review on Machine Learning, Big Data Analytics, and Design for Additive Manufacturing for Aerospace Applications,"

Journal of Materials Engineering and Performance, vol. 31, no. 8, pp. 6112-6130, 2022/08/01 2022.

[169] P. Ezanno et al., "Research perspectives on animal health in the era of artificial intelligence," *Veterinary Research,* vol. 52, no. 1, p. 40, 2021/03/06 2021.

[170] D. Ververidis et al., "An authoring tool for democratizing the creation of high-quality VR experiences," *Virtual Reality,* vol. 26, no. 1, pp. 105-124, 2022/03/01 2022.

[171] I. H. J. S. c. s. Sarker, "*Investigating undergraduate students' intention to engage in artificial intelligence learning*," vol. 3, no. 2, pp. 158-158.

[172] A. Tolk, P. Barry, and S. C. Doskey, "Using modeling and simulation and artificial intelligence to improve complex adaptive systems engineering," *International Journal of Modeling, Simulation, and Scientific Computing,* vol. 13, no. 02, p. 2241004, 2022/04/01 2021.

[173] K. Choudhary et al., "Recent advances and applications of deep learning methods in materials science," *npj Computational Materials,* vol. 8, no. 1, p. 59, 2022/04/05 2022.

[174] J. P. Lai, Y. L. Lin, H. C. Lin, C. Y. Shih, Y. P. Wang, and P. F. Pai, "T*ree-Based Machine Learning Models with Optuna in Predicting Impedance Values for Circuit Analysis*," vol. 14, no. 2, p. 265, 2023.

[175] S. Masoumi-Verki, F. Haghighat, and U. Eicker, "A review of advances towards efficient reduced-order models (ROM) for predicting urban airflow and pollutant dispersion," *Building and Environment,* vol. 216, p. 108966, 2022/05/15/ 2022.

[176] S. S. Babu, A.-H. I. Mourad, K. H. Harib, and S. Vijayavenkataraman, "Recent developments in the application of machine-leaning towards accelerated predictive multiscale design and additive manufacturing," *Virtual and Physical Prototyping,* vol. 18, no. 1, p. e2141653, 2023/01/01 2023.

[177] S. Sai, R. Sai, and V. Chamola, "Generative AI for Industry 5.0: Analyzing the impact of ChatGPT, DALLE, and Other Models," *IEEE Open Journal of the Communications Society,* pp. 1-1, 2024.

[178] S. J. Zaidi and H. Saleem, *Reverse Osmosis Systems: Design, Optimization and Troubleshooting Guide*. Elsevier, 2021.

[179] S. Sundaramoorthy, G. Srinivasan, and D. V. R. Murthy, "An analytical model for spiral wound reverse osmosis membrane modules: Part I — Model development and parameter estimation," *Desalination,* vol. 280, no. 1, pp. 403-411, 2011/10/03/ 2011.

[180] M. A. Al-Obaidi, A. A. Alsarayreh, A. Bdour, S. H. Jassam, F. L. Rashid, and I. M. Mujtaba, "Simulation and optimisation of a medium scale reverse osmosis brackish water desalination system under variable feed quality: Energy saving and maintenance opportunity," *Desalination,* vol. 565, p. 116831, 2023/11/01/ 2023.

[181] C. Chen and H. Qin, "*A Mathematical Modeling of the Reverse Osmosis Concentration Process of a Glucose Solution*," vol. 7, no. 5, p. 271, 2019.

[182] S. A., L. A. H., and S. I., "Simulation of A Reverse Osmosis Seawater Desalination Plant, Part 1: Theoretical Modeling of A Real Plant and Its Validation %J the International Conference on Applied Mechanics and Mechanical Engineering,"

vol. 13, no. *13th International Conference on Applied Mechanics and Mechanical Engineering.*, pp. 464-482, 2008.

[183] K. T. Chua, M. N. A. Hawlader, and A. Malek, "Pretreatment of seawater: Results of pilot trials in Singapore," *Desalination,* vol. 159, no. 3, pp. 225-243, 2003/11/05/ 2003.

[184] "Index," in *Membranes for Clean and Renewable Power Applications*, A. Gugliuzza and A. Basile, Eds.: Woodhead Publishing, 2014, pp. 397-410.

[185] K. Poirier et al., "*A comprehensive review of pre-and post-treatment approaches to achieve sustainable desalination for different water streams*," vol. 566, p. 116944, 2023.

[186] S. Jamaly, N. Darwish, I. Ahmed, and S. J. D. Hasan, "*A short review on reverse osmosis pretreatment technologies*," vol. 354, pp. 30-38, 2014.

[187] S. Jiang, Y. Li, and B. P. J. S. o. t. t. e. Ladewig, "*A review of reverse osmosis membrane fouling and control strategies*," vol. 595, pp. 567-583, 2017.

[188] N. J. D. Voutchkov, "*Considerations for selection of seawater filtration pretreatment system,*" vol. 261, no. 3, pp. 354-364, 2010.

[189] A. Giwa, V. Dufour, F. Al Marzooqi, M. Al Kaabi, and S. J. D. Hasan, "*Brine management methods: Recent innovations and current status,*" vol. 407, pp. 1-23, 2017.

[190] L. Y. Lee et al., "*Ozone-biological activated carbon as a pretreatment process for reverse osmosis brine treatment and recovery*," vol. 43, no. 16, pp. 3948-3955, 2009.

[191] W. Gao et al., "*Membrane fouling control in ultrafiltration technology for drinking water production: A review*," vol. 272, no. 1-3, pp. 1-8, 2011.

[192] N. M. Farhat, C. Christodoulou, P. Placotas, B. Blankert, O. Sallangos, and J. S. J. D. Vrouwenvelder, "*Cartridge filter selection and replacement: optimization of produced water quantity, quality, and cost,*" vol. 473, p. 114172, 2020.

[193] F. E. Ahmed, R. Hashaikeh, A. Diabat, and N. J. D. Hilal, "*Mathematical and optimization modelling in desalination: State-of-the-art and future direction*," vol. 469, p. 114092, 2019.

[194] M. M. Zubair, H. Saleem, and S. J. Zaidi, "Recent progress in reverse osmosis modeling: An overview," *Desalination,* vol. 564, p. 116705, 2023/10/15/ 2023.

[195] V. G. J. D. Gude and w. treatment, "*Energy consumption and recovery in reverse osmosis*," vol. 36, no. 1-3, pp. 239-260, 2011.

[196] J. Johnson and M. Busch, "Engineering Aspects of Reverse Osmosis Module Design," *Desalination and Water Treatment,* vol. 15, no. 1-3, pp. 236-248, 2010/03/01 2010.

[197] N. Prihasto, Q.-F. Liu, and S.-H. Kim, "Pre-treatment strategies for seawater desalination by reverse osmosis system," *Desalination,* vol. 249, no. 1, pp. 308-316, 2009/11/30/ 2009.

[198] C. Koutsou, S. Yiantsios, and A. J. J. o. m. s. Karabelas, "*Direct numerical simulation of flow in spacer-filled channels: Effect of spacer geometrical characteristics*," vol. 291, no. 1-2, pp. 53-69, 2007.

[199] G. Fimbres-Weihs, D. J. C. E. Wiley, and P. P. Intensification, "R*eview of 3D CFD modeling of flow and mass transfer in narrow spacer-filled channels in membrane modules,*" vol. 49, no. 7, pp. 759-781, 2010.

[200] C. P. Koutsou, S. G. Yiantsios, and A. J. J. J. o. M. S. Karabelas, "A nu*merical and experimental study of mass transfer in spacer-filled channels: Effects of spacer geometrical characteristics and Schmidt number,*" vol. 326, no. 1, pp. 234-251, 2009.

[201] S. A, L. A. H, and S. I, "Simulation of a Reverse Osmosis Seawater Desalination Plant, Part 2: Experimental Measurements and Prediction Study on the Validated Simulation Model," (in en), *The International Conference on Applied Mechanics and Mechanical Engineering,* vol. 13, no. 13th International Conference on Applied Mechanics and Mechanical Engineering., pp. 483-505, 2008.

[202] L. Cheng and T. J. I. J. o. E. R. Yu, "*A new generation of AI: A review and perspective on machine learning technologies applied to smart energy and electric power systems*," vol. 43, no. 6, pp. 1928-1973, 2019.

[203] S. I. Abba et al., "Optimizing sustainable desalination plants with advanced ML-based uncertainty analysis," *Applied Soft Computing,* vol. 169, p. 112624, 2025/01/01/ 2025.

[204] S. Safeer et al., "A review of artificial intelligence in water purification and wastewater treatment: Recent advancements," *Journal of Water Process Engineering,* vol. 49, p. 102974, 2022/10/01/ 2022.

[205] A. R. Habieeb, A. E. Kabeel, G. I. Sultan, and M. M. Abdelsalam, "Advancements in Water Desalination Through Artificial Intelligence: A Comprehensive Review of AI-Based Methods for Reverse Osmosis Membrane Processes," *Water Conservation Science and Engineering,* vol. 8, no. 1, p. 53, 2023/10/13 2023.

[206] R. Shahouni, M. Abbasi, M. Dibaj, and M. Akrami, "*Utilising Artificial Intelligence to Predict Membrane Behaviour in Water Purification and Desalination,*" vol. 16, no. 20, p. 2940, 2024.

[207] R. Mahadeva, G. Manik, A. Goel, and N. Dhakal, "A review of the artificial neural network-based modelling and simulation approaches applied to optimize reverse osmosis desalination techniques," *Desalination and Water Treatment,* vol. 156, pp. 245-256, 2019/07/01/ 2019.

[208] R. Mahadeva, G. Manik, O. P. Verma, and S. Sinha, "Modelling and simulation of desalination process using artificial neural network: a review," *Desalination and Water Treatment,* vol. 122, pp. 351-364, 2018/08/01/ 2018.

[209] S. U.-D. Khan and J. Orfi, "*Socio-Economic and Environmental Impact of Nuclear Desalination,*" vol. 13, no. 12, p. 1637, 2021.

[210] M. Soori, F. K. G. Jough, R. Dastres, and B. Arezoo, "AI-Based Decision Support Systems in Industry 4.0, A Review," *Journal of Economy and Technology,* 2024/08/28/ 2024.

[211] S. Rani, A. Kataria, and M. Chauhan, "Cyber security techniques, architectures, and design," in *Holistic approach to quantum cryptography in cyber security*: CRC Press, 2022, pp. 41-66.

[212] A. Shafiei, A. Tatar, M. Rayhani, M. Kairat, and I. Askarova, "Artificial neural network, support vector machine, decision tree, random forest, and committee

machine intelligent system help to improve performance prediction of low salinity water injection in carbonate oil reservoirs," *Journal of Petroleum Science and Engineering,* vol. 219, p. 111046, 2022/12/01/ 2022.

[213] S. F. Ahmed et al., "Deep learning modelling techniques: current progress, applications, advantages, and challenges," *Artificial Intelligence Review,* vol. 56, no. 11, pp. 13521-13617, 2023/11/01 2023.

[214] A. G. J. A. o. c. m. i. e. Gad, "*Particle swarm optimization algorithm and its applications: a systematic review*," vol. 29, no. 5, pp. 2531-2561, 2022.

[215] A. K. Kesarwani, M. Yadav, D. Singh, and G. D. Gautam, "A review on the recent applications of particle swarm optimization & genetic algorithm during antenna design," *Materials Today: Proceedings,* vol. 56, pp. 3823-3825, 2022/01/01/ 2022.

[216] A. M. A. J. A. U. Fadul, "*Anomaly Detection based on Isolation Forest and Local Outlier Factor*," 2023.

[217] D. Divya, B. Marath, and M. J. J. o. Q. i. M. E. Santosh Kumar, "*Review of fault detection techniques for predictive maintenance*," vol. 29, no. 2, pp. 420-441, 2023.

[218] A. A. Khan and J. J. S. Abonyi, "*Simulation of sustainable manufacturing solutions: Tools for enabling circular economy*," vol. 14, no. 15, p. 9796, 2022.

[219] S. P. Langarudi, R. P. Sabie, B. Bahaddin, and A. G. J. M. Fernald, "*A Literature Review of Hybrid System Dynamics and Agent-Based Modeling in a Produced Water Management Context,*" vol. 2, no. 2, pp. 224-239, 2021.

[220] K. Agalianos, S. Ponis, E. Aretoulaki, G. Plakas, and O. J. P. M. Efthymiou, "*Discrete event simulation and digital twins: review and challenges for logistics,*" vol. 51, pp. 1636-1641, 2020.

[221] F. E. Cellier and J. Greifeneder, *Continuous system modeling*. Springer Science & Business Media, 2013.

[222] S. C. Brailsford, T. Eldabi, M. Kunc, N. Mustafee, and A. F. J. E. J. o. O. R. Osorio, "*Hybrid simulation modelling in operational research: A state-of-the-art review,*" vol. 278, no. 3, pp. 721-737, 2019.

[223] J. Mun, *Modeling risk: Applying Monte Carlo simulation, real options analysis, forecasting, and optimization techniques*. John Wiley & Sons, 2006.

[224] M. Lippe et al., "Using agent-based modelling to simulate social-ecological systems across scales," vol. 23, no. 2, pp. 269-298, 2019.

[225] R. Sabharwal, *Introduction to System Dynamics*. Educohack Press, 2025.

[226] S. V. J. W. E. Berg and Policy, "*Performance assessment using key performance indicators (KPIs) for water utilities: a primer*," vol. 6, no. 02, p. 2050001, 2020.

[227] G. Rediske, L. Michels, J. Cezar Mairesse Siluk, P. Donaduzzi Rigo, C. Brum Rosa, and R. J. I. J. o. E. R. Jochann Franceschi Bortolini, "*Management of operation and maintenance practices in photovoltaic plants: Key performance indicators,*" vol. 46, no. 6, pp. 7118-7136, 2022.

[228] K. Mahmood, T. Otto, E. Shevtshenko, and T. Karaulova, "Performance evaluation by using overall equipment effectiveness (OEE): an analyzing tool," in *International conference on innovative technologies*, 2016, vol. 2016, pp. 6-8.

[229] S. Duer et al., "R*eliability testing of wind farm devices based on the mean time between failures (MTBF),*" vol. 16, no. 4, p. 1659, 2023.

[230] P. E. J. J. S. Kabore and Transducers, "*Simulating Specific Energy Consumption for an Industrial Production Process*," vol. 265, no. 2, pp. 40-47, 2024.

[231] S. Chidiac, P. El Najjar, N. Ouaini, Y. El Rayess, D. J. R. i. E. S. El Azzi, and Bio/Technology, "*A comprehensive review of water quality indices (WQIs): history, models, attempts and perspectives,*" vol. 22, no. 2, pp. 349-395, 2023.

[232] J. García, A. Leiva-Araos, E. Diaz-Saavedra, P. Moraga, H. Pinto, and V. J. A. S. Yepes, "*Relevance of Machine Learning Techniques in Water Infrastructure Integrity and Quality: A Review Powered by Natural Language Processing*," vol. 13, no. 22, p. 12497, 2023.

[233] H. S. Kang et al., "Optimization of pumping schedule based on water demand forecasting using a combined model of autoregressive integrated moving average and exponential smoothing," vol. 15, no. 1, pp. 188-195, 2015.

[234] C. Ortiz-Lopez, C. Bouchard, and M. J. E. T. R. Rodriguez, "Machine learning models with potential application to predict source water quality for treatment purposes: a critical review," vol. 11, no. 1, pp. 118-147, 2022.

[235] M. Dastorani, M. Mirzavand, M. T. Dastorani, and H. J. S. W. R. M. Khosravi, "*Simulation and prediction of surface water quality using stochastic models*," vol. 6, pp. 1-17, 2020.

[236] Z. Dong, L. Chuan, L. Yongxiang, G. J. Q. Zhiqi, and R. E. International, "*A system's mean time to repair allocation method based on the time factors*," vol. 30, no. 2, pp. 247-256, 2014.

[237] F. S. Pinto and R. C. Marques, "Desalination projects economic feasibility: A standardization of cost determinants," *Renewable and Sustainable Energy Reviews,* vol. 78, pp. 904-915, 2017/10/01/ 2017.

[238] E. Edozie, A. N. Shuaibu, B. O. Sadiq, and U. K. John, "Artificial intelligence advances in anomaly detection for telecom networks," *Artificial Intelligence Review,* vol. 58, no. 4, p. 100, 2025/01/25 2025.

[239] S. O. Pinto and V. A. Sobreiro, "Literature review: Anomaly detection approaches on digital business financial systems," *Digital Business,* vol. 2, no. 2, p. 100038, 2022/01/01/ 2022.

[240] N. Najid et al., "Fouling control and modeling in reverse osmosis for seawater desalination: A review," *Computers & Chemical Engineering,* vol. 162, p. 107794, 2022/06/01/ 2022.

[241] A. Giwa, H. Ademola, and A. O. Yusuf, "Machine learning and application for modeling and prediction of desalination cost globally," *Desalination,* vol. 608, p. 118829, 2025/08/01/ 2025.

[242] M. S. Ismail, A. M. Mohamed, D. Poggio, M. Walker, and M. Pourkashanian, "Modelling mass transport within the membrane of direct contact membrane distillation modules used for desalination and wastewater treatment: Scrutinising assumptions," *Journal of Water Process Engineering,* vol. 45, p. 102460, 2022/02/01/ 2022.

[243] Y. K. Chong, Y. Y. Liang, W. J. Lau, and G. A. Fimbres Weihs, "3D CFD study of hydrodynamics and mass transfer phenomena for spiral wound membrane submerged-type feed spacer with different node geometries and sizes,"

International Journal of Heat and Mass Transfer, vol. 191, p. 122819, 2022/08/01/ 2022.

[244] Y. Y. Liang, "Review of analytical and numerical modeling for pressure retarded osmosis membrane systems," *Desalination,* vol. 560, p. 116655, 2023/08/15/ 2023.

[245] F. Mohammadi, S.-A. Mostafa, A.-A. Yousef, and G. T. and Heydt, "Cost-Benefit Analysis of Desalination: A Power Market Opportunity," *Electric Power Components and Systems,* vol. 48, no. 11, pp. 1091-1101, 2020/07/02 2020.

[246] S. S and S. S, "Reverse Osmosis–Pressure Retarded Osmosis hybrid system: Modelling, simulation and optimization," *Desalination,* vol. 389, pp. 78-97, 2016/07/01/ 2016.

[247] M. S. Shalaby et al., "Economic assessment for TFC-RO membranes production for water desalination," *Chemical Engineering Science,* vol. 288, p. 119805, 2024/04/15/ 2024.

[248] A. Aljuwaisseri, E. Aleisa, and K. Alshayji, "Environmental and economic analysis for desalinating seawater of high salinity using reverse osmosis: a life cycle assessment approach," *Environment, Development and Sustainability,* vol. 25, no. 5, pp. 4539-4574, 2023/05/01 2023.

[249] R. Colciaghi, R. Simonetti, L. Molinaroli, M. Binotti, and G. Manzolini, "Levelized cost of water assessment for small-scale desalination plant based on forward osmosis process," *Energy Conversion and Management,* vol. 271, p. 116336, 2022/11/01/ 2022.

[250] M. Smaoui and L. Krichen, "Control, energy management and performance evaluation of desalination unit based renewable energies using a graphical user interface," *Energy,* vol. 114, pp. 1187-1206, 2016/11/01/ 2016.

About the Author

Amr Seoudy has more than 30 years' experience in Operation and Maintenance, Research and Development, Process Design, Basic and Detailed Engineering, Integrated Projects Management, Technical and Financial Studies, Events Organization and Training, and in the field of *Water Treatment Systems* and *Water Infrastructure Facilities*.

He as an MSc in "Simulating of Reverse Osmosis Seawater Desalination Systems."

Had activities in the field of training and research, (13) training courses. Participated in many local and international seminars, conferences and congresses with (22) lectures, (20) papers.

Participated and led (323) projects in the field of *Water Treatment Systems and Infrastructure Facilities*, whether studies, designs, supervision or project management.

Published recently in high ranked journals like "Desalination and water treatment" Q3 Scopus and "Results in Engineering" Q1 Scopus.

Writer for a Perspective Chapter: "*Desalination System Performance Index and Artificial Intelligence*," Advances in Desalination Insights, Aug. 22, 2024.

Amr Seoudy, Founder & CEO of WTCO Consulting and Engineering firm from 2012.

In 2019, established the first branch of the office outside Egypt, in Dubai - United Arab Emirates, to provide office's activities in Arabian Gulf region and Middle East zones.

Amr Seoudy, Founder & Organizer for "*International Water Consultant Congress (IWCC)*" from 2020 and Founder of "*Amr Seoudy Water Treatment Award (ASWTA)*."

In 2022, established the second branch of the office outside Egypt, in Oran - People's Democratic Republic of Algeria, with cooperation with other partners, to provide office's activities in North, West and Central Africa.

Hatem Seoudy, Received his B.Sc. Higher Institute of Engineering, El-Shorouk Academy, Egypt, in 2005, M.Sc. degree Faculty of Engineering, Helwan University, Egypt, in 2010, and Ph.D. Faculty of Engineering, Al-Azhar University, Egypt, in 2016. Assoc. Prof. in 2024. Currently an Assoc. Prof. at Higher Institute of Engineering, El-Shorouk Academy.

Index

E

F

H

I

L

M

O

P

R

S

T

W